ELECTRONICS DRAFTING WORKBOOK

Third Edition

Cyrus Kirshner, Ph.D.

Professor of Engineering
Los Angeles Valley College

Kurt M. Stone

Senior Engineering Designer
ITT Gilfillan

Instructor, Electronics Drafting
West Valley Occupational Center

McGRAW-HILL BOOK COMPANY, Gregg Division

New York	Johannesburg	Paris
St. Louis	London	São Paulo
Dallas	Madrid	Singapore
San Francisco	Mexico	Sydney
Auckland	Montreal	Tokyo
Bogotá	New Delhi	Toronto
Düsseldorf	Panama	

Equipment Used in Electronics Drafting

The equipment used in electronics drafting is, for the most part, the same as that used in mechanical drafting, such as:

Drawing board
24″ T square
45°/90° triangle
30°/60° triangle
Compass
X-acto knife (for printed circuit tapes)
Protractor
Scale (12″ preferred with both
    1/32″ and 1/50″ increments)
French curve
Erasing shield
Dusting brush
Drafting (or masking) tape
Ruby drafting eraser
Drafting pencils (H, 2H, and 4H) automatic preferred
Sandpaper pad or file

The above list is the usual equipment required for mechanical drafting, which is the same as required for electronics drafting. The student having these items but in different sizes need not purchase any new equipment other than electronics templates:

Electronics template USAS Y32.2
Logic template MIL-STD-806 (Y32.14) three-quarter size

Library of Congress Cataloging in Publication Data

Kirshner, Cyrus.
   Electronics drafting workbook.

   1. Electronic drafting. I. Stone, Kurt M.,
     joint author. II. Title
TK7866.K57   1977   604'.2'621381   77-7872
ISBN 0-07-034890-1

ELECTRONICS DRAFTING WORKBOOK, Third Edition

7890EBEB 832

We respectfully dedicate this book to our families and our friends, and to the advancement of technical education.

Contents

Preface

The third edition of the **Electronics Drafting Workbook** was written in the same format and style of the very successful first two editions. Our main purpose in the third edition is to update the latest practices being adopted by industry.

The new updated topics being covered are those dealing with design and layout of dual-in-line integrated circuits. Along with new lessons explaining the new practice, we have introduced accompanying appendixes which will be needed to complete the new problems.

The book is written for the student between high school and college, say the last two years of high school or the first two years of college. It can also be used easily by adult education programs as well as by correspondence or home-study students. The material presented is largely based on the electronics drafting course taught at Los Angeles Valley College since 1959 for both day and evening classes, as well as at the West Valley Occupational Center, Woodland Hills, California, with a great deal of success. The course at Los Angeles Valley College is credit-transferable to California Polytechnic Institute (which is fully accredited) and meets the electronics drafting requirements for the electronics engineer's degree.

The sequence of lessons was designed with these considerations in mind:

1. To progress from a simpler, or more familiar, kind of drawing to one that is slightly more complex. Therefore it would be well to follow the exercises in order where feasible.

2. To start the student out immediately doing electronically related drafting, without spending too much time on conventional mechanical drafting review, and to develop a consecutive interest in electronics drafting concepts.

3. To present the most frequently used types of drawings first and infrequently used types of drawings last (i.e., based on the experiences of most beginning draftsmen).

4. To culminate the various ideas in a project at the end of the semester, or at least to indicate how this can be done.

The basic technique of instruction used here is to present a little theory and an exercise on each page with a step-by-step "learn by doing" approach. Another technique is to develop the habit of looking up new information (in the appendix, catalogs, etc.) and checking back old information (from previous exercises, drawings, etc.).

The authors wish to thank the many instructors, engineers, drafting supervisors, and companies who contributed to the preparation of this book. In particular, we would like to mention:

James J. Bots, Meramec Community College, St. Louis, Missouri.

ITT Gilfillan, San Fernando, California

The General Electric Company, for the FM Tuner schematic diagram (page 27) and the Power Supply schematic diagram (page 55)

RCA, for the color and black and white TV Receiver block diagram (page 20)

The Zenith Electric Co. for the color TV block diagram (page 22)

We would also like to express sincere appreciation to the following firms which contributed to, and are represented in, our appendixes:

Arco Electronics, Inc., Alpha Wire Corp., General Electric Company, Grayhill, Inc., Herman H. Smith, Inc., Hughes Semiconductors, Littelfuse Incorporated, Ohmite Mfg. Co., Texas Instruments Incorporated, Triad Transformer Corporation, USECO, Vitramon, Inc., Westinghouse Electric Corporation, and Department of Defense Military Standards.

Cyrus Kirshner
Kurt Stone

Suggestions to the Instructor

A Solutions Manual has been prepared for the instructor who may not have enough time to work out each solution by himself but who would nevertheless like an answer to each problem. It should be pointed out, however, that there are a few problems which can have more than one solution which is also emphasized in the Solutions Manual.

The Solutions Manual is available only to instructors from the publisher upon request.

Electromechanical **lettering** need not be artistic, but it should be neat and legible.

Copy the **notes** taken from column 1 started below, going from bottom to top.

Copy in column 4 the entries of column 3 using ⅛" high vertical lettering.

NOTES:

1) ELECTRONICS LET-TERING IS LARGELY DONE FREEHAND ON VELLUM WITH OR AGAINST A 1/8 OR 1/10 GRID BACKGROUND.

2) THE LETTERING CAN BE EITHER VERTICAL OR *INCLINED*.

3) THE STUDENT SHOULD LEARN BOTH STYLES. HOWEVER, IT IS EASIER TO LEARN VERTICAL LETTERING FIRST AND *INCLINED LETTERING 2ND*, RATHER THAN VICE VERSA.

4) PRACTICE DEVELOPS TECHNIQUE.

4) PRACTICE

1) ELECTRONICS LET-TERING IS LARGELY

NOTES:

$2\frac{1}{2}$ $3\frac{3}{4}$ $4\frac{5}{64}$ $5\frac{7}{64}$

$6\frac{7}{8}$ $7\frac{9}{32}$ $8\frac{9}{16}$ $9\frac{1}{10}$

$5.110 \overset{+.005}{}$ DIA. $6.120 \overset{+.002}{-.000}$

TOL.= TOLERANCE

FRAC.= ± 1/64; X°= ± 1/2°

.XX= ± .03; .XXX= ± .010

2.250
2.251

CLEARANCE FIT

2.248
2.247

2.2519
2.2513

INTERFERENCE FIT

2.2500
2.2506

TITLE	MECHANICAL DRAFTING REVIEW—LETTERING		DWG. NO. **MDR-1**	PAGE 1
NAME	DATE	COURSE	GRADE	SCALE	SHEET 1 OF 2

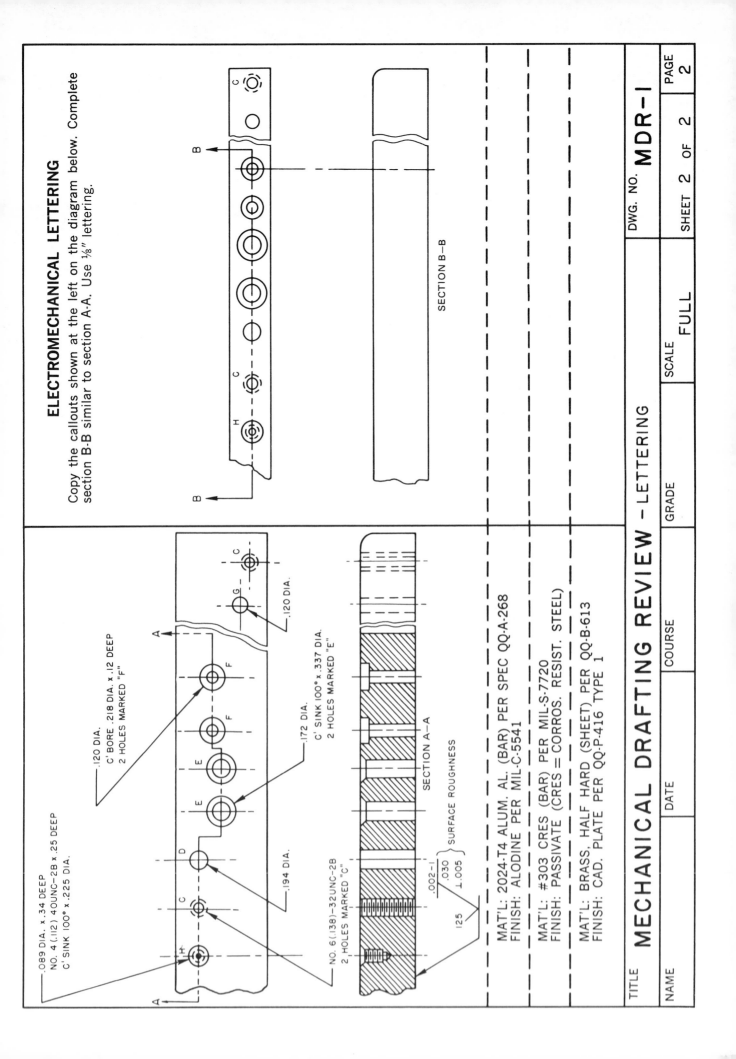

ELECTROMECHANICAL LETTERING

Copy the callouts shown at the left on the diagram below. Complete section B-B similar to section A-A. Use ⅛" lettering.

SECTION B—B

.089 DIA. x .34 DEEP
NO. 4 (.112) 40UNC–2B x .25 DEEP
C' SINK 100° x .225 DIA.

.120 DIA.

C' BORE .218 DIA. x .12 DEEP
2 HOLES MARKED "F"

.172 DIA.

C' SINK 100° x .337 DIA.
2 HOLES MARKED "E"

.194 DIA.

NO. 6 (.138)–32UNC–2B
2 HOLES MARKED "C"

.120 DIA.

.120 DIA.

SECTION A—A

.002–1
.030
⊥.005 } SURFACE ROUGHNESS

.125

MAT'L: 2024-T4 ALUM. AL. (BAR) PER SPEC QQ-A-268
FINISH: ALODINE PER MIL-C-5541

MAT'L: #303 CRES (BAR) PER MIL-S-7720
FINISH: PASSIVATE (CRES = CORROS. RESIST. STEEL)

MAT'L: BRASS, HALF HARD (SHEET) PER QQ-B-613
FINISH: CAD. PLATE PER QQ-P-416 TYPE 1

TITLE	MECHANICAL DRAFTING REVIEW – LETTERING		DWG. NO. MDR–1	
			SCALE FULL	PAGE 2
NAME	DATE	COURSE	GRADE	SHEET 2 OF 2

DIMENSIONING AND LINE WEIGHTS

VISIBLE LINE: thick (H lead)
For outline and cutting planes.

HIDDEN LINE: medium (2H lead)

SECTION LINE: thin (4H lead)
For center lines, section, dimension, extension, and phantom lines.

With a little practice, all lines mentioned above can be made with a 2H lead by varying pressure and relining.

PANEL FRONT VIEW

Exercise. The panel above is only partially **dimensioned** at half scale. Redraw the panel full scale and dimension it completely. Omit section A-A.

Hint: The missing dimensions will have to be found by **scaling** the half scale drawing above and **multiplying by 2.**

Use your template for all holes and radii.

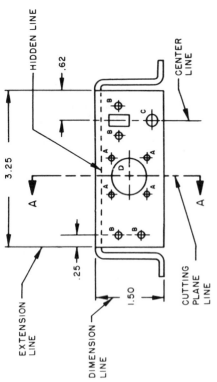

EXTENSION LINE

HIDDEN LINE

3.25

.62

DIMENSION LINE

.25

1.50

CUTTING PLANE LINE

CENTER LINE

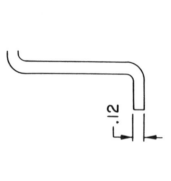

1.50

.38

.38

.19

SECTION LINES

SECTION A-A

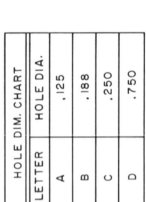

.12

HOLE DIM. CHART	
LETTER	HOLE DIA.
A	.125
B	.188
C	.250
D	.750

DIMENSIONING FROM A DATUM

On sheet 1 you used the **conventional double arrow and dimension line** method, shown again in Fig. 1. If space is limited, **arrowless dimensioning from a datum** is preferred, as shown in Fig. 2.

Avoid dimensioning to hidden lines. Dimension another view. **A sectional view** may be necessary.

Use **decimal dimensioning** for all work.

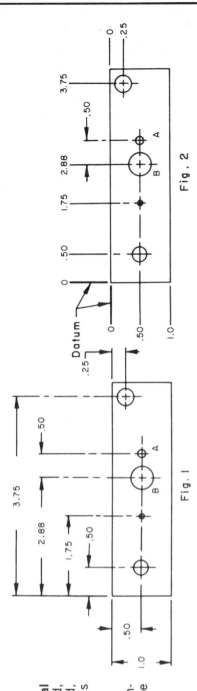

Fig. 1

Fig. 2

Note: Location of hole A is critical to hole B; therefore the dimension is direct to hole B and not to the datum.

Exercise. Redraw the front and section A-A of the panel on sheet 1 using the **datum dimensioning** method, full scale.

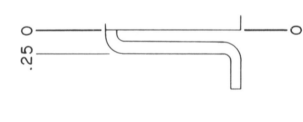

TITLE MECHANICAL DRAFTING REVIEW—DIMENSIONING, DATUM   DWG. NO. MDR-2   PAGE 4

NAME	DATE	COURSE	GRADE	SCALE	DWG. NO. MDR-2
				NOTED	SHEET 2 OF 3

DIMENSIONING

The drawing below is a typical partially dimensioned example showing the use of standard hole layouts.

Exercise. Dimension the missing hole locations and draw the hole pattern properly in place. Use the **datum dimensioning method** (arrowless dimensioning). Scale: half size. Lettering ⅛".

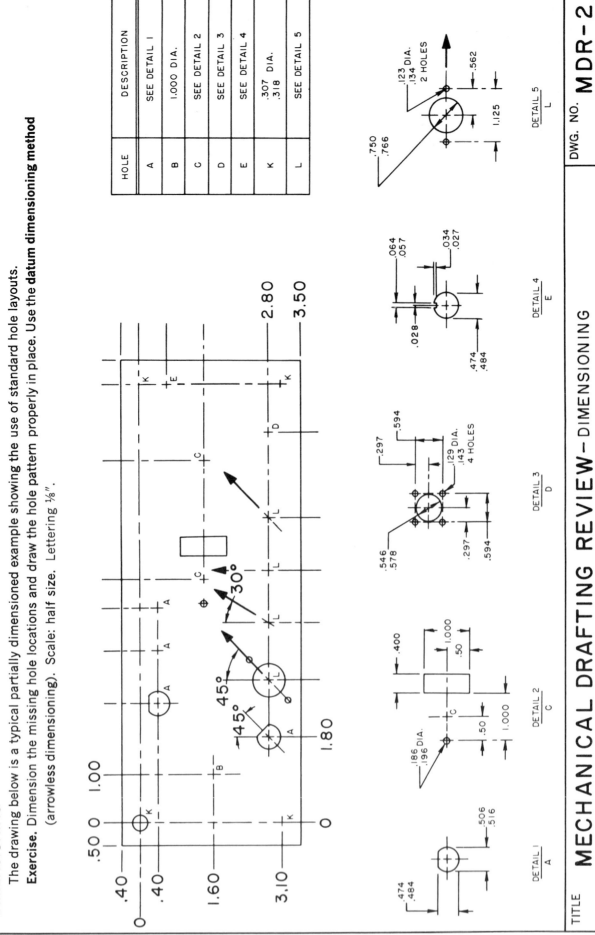

HOLE	DESCRIPTION
A	SEE DETAIL 1
B	1.000 DIA.
C	SEE DETAIL 2
D	SEE DETAIL 3
E	SEE DETAIL 4
K	.307 DIA. .318
L	SEE DETAIL 5

DETAIL 5
L

DETAIL 4
E

DETAIL 3
D

DETAIL 2
C

DETAIL 1
A

TITLE	MECHANICAL DRAFTING REVIEW – DIMENSIONING		DWG. NO. MDR-2	
NAME	DATE	COURSE	GRADE	SCALE HALF SIZE	SHEET 3 OF 3

MILITARY STANDARDS (MS)

Military Standards become known to draftsmen by many code numbers. Examples are:

MIL-STD-15 (electrical and electronics symbols)

MIL-A-8625 (anodize)

QPL-641-16 (jacks, telephone)

JAN-S-28 (sockets, electronic tube)

QQ-A-327 (aluminum sheet and plate)

MS35221-45 (screw, pan head)

In addition, there are hundreds whose code letters start with AN-, ANA-, EI-, AND-, and many others. These various Military Standards permit the electromechanical draftsman to use a **short code number** to represent a lengthy description or detailed drawing.

In industry the electronics draftsman spends a great deal of his time looking up electronic components, hardware and specs. Basically, the MS (Military Standard) number describes one of the following:

A booklet of general information

An engineering material

A fabricating process

An electronic component

A piece of hardware

Handling procedure (assembly, installation, or operation of equipment)

See examples of typical pages of Military Standards in Appendix C, pages 97-100.

Exercises. In order to answer the exercises below, refer to Appendix C, pages 97-100 (Military Standards).

1. Complete MS35221-14 callout below with its material and finish.

#4-40NC X 5/16 LONG, PAN HEAD SCREW

Material _____ Finish _____

2. What is MS35649-84? _____

3. What is the outside diameter of MS35337-3? _____

4. What is the MS number of the following:

#2, WASHER, FLAT GEN. PURPOSE CRES _____

#2, WASHER, FLAT GEN. PURPOSE BRASS _____

#6, LOCK WASHER, SPLIT, LIGHT SERIES, PLAIN _____

---

5. What information is given in the first four columns of MS15795? _____

What are the rest of the columns used for? _____

6. What are the MS numbers of all the hardware shown in Fig. 1?

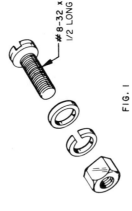

#8-32 x 1/2 LONG

(HARDWARE MATERIAL — CRES)

FIG. 1

SCREW _____

NUT _____

WASHER, FLAT _____

WASHER, LOCK _____

TITLE	MILITARY STANDARDS—INTRODUCTION			DWG. NO. MS-1	
NAME	DATE	COURSE	GRADE	SCALE NONE	PAGE 34
				SHEET 1 OF 2	

A typical callout of a fastening using MS numbers might read as follows;

Example

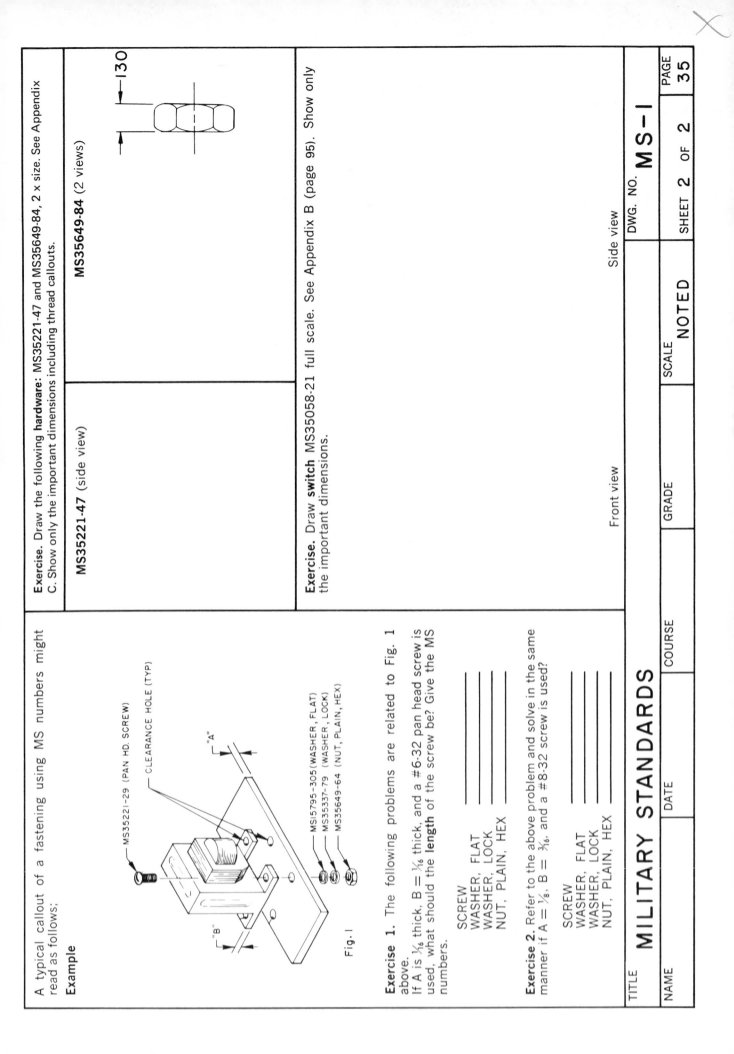

MS35221-29 (PAN HD. SCREW)
CLEARANCE HOLE (TYP)
"A"
"B"

MS15795-305 (WASHER, FLAT)
MS35337-79 (WASHER, LOCK)
MS35649-64 (NUT, PLAIN, HEX)

Fig. I

Exercise 1. The following problems are related to Fig. 1 above.
If A is $\frac{1}{16}$ thick, B = $\frac{1}{16}$ thick, and a #6-32 pan head screw is used, what should the **length** of the screw be? Give the MS numbers.

SCREW _____
WASHER, FLAT _____
WASHER, LOCK _____
NUT, PLAIN, HEX _____

Exercise 2. Refer to the above problem and solve in the same manner if A = $\frac{1}{8}$, B = $\frac{3}{16}$, and a #8-32 screw is used?

SCREW _____
WASHER, FLAT _____
WASHER, LOCK _____
NUT, PLAIN, HEX _____

Exercise. Draw the following **hardware**: MS35221-47 and MS35649-84, 2 x size. See Appendix C. Show only the important dimensions including thread callouts.

MS35649-84 (2 views)

.130

MS35221-47 (side view)

Exercise. Draw **switch** MS35058-21 full scale. See Appendix B (page 95). Show only the important dimensions.

Front view   Side view

PRINTED CIRCUIT PATTERN

The length of the conductors between various lands shall be held to a minimum.

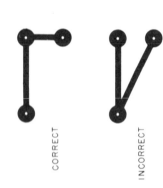

CORRECT

INCORRECT

CORRECT

LAND

CONDUCTOR

INCORRECT

Exercise 1: Connect the lands given in Figs. 1, 2, and 3 with a $\frac{1}{16}$" wide black pressure-sensitive tape. Leave $\frac{1}{16}$" minimum spacing from edge of board and component holes. If no tape is available, use pencil and draw the conductors $\frac{1}{16}$" wide (A tee, see Fig. 4, with fillets is optional.)

Fig. 1

Fig. 2

Fig. 3

Exercise 2. With a $\frac{1}{16}$" wide tape connect the following in Fig. 4.

All lands marked A join together.
All lands marked B join together.
Connect nearest A hole to remaining connector tab (as shown with B hole to tab).

Make all connections on one side of board only, as shown, and keep the length of the conductors to a minimum.

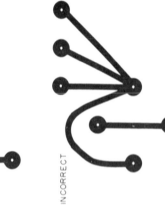

B
B
B
A
A
B
A
A
B
B
B

TEE

CONNECTOR TAB

KEY

Fig. 4

Exercise 3. With a $\frac{1}{16}$" wide tape connect the following in Fig. 5:

All 1 lands together to **Grd** (ground).
All 2 lands together to **Out**.
All 6 lands together.

Connect transistor as follows:
Base B to No. 4 tab.
Collector C to No. 5 tab.
Emitter E to land 3.

Avoid sharp corners.

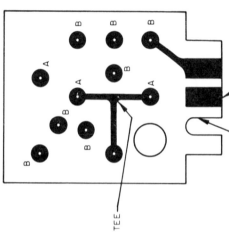

SPARE
NO. 4
SPARE
NO. 5
SPARE
OUT
SPARE
GRD

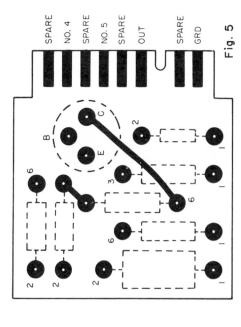

Fig. 5

TITLE	PRINTED CIRCUIT BOARD–PRINTED CIRCUIT PATTERN		DWG. NO. PCB–1	
NAME	DATE	COURSE	GRADE	SCALE 2 X SIZE
				PAGE 36
				SHEET 1 OF 2

TRANSISTOR AND DIODES IN PRINTED CIRCUIT BOARD (PCB)

Special attention should be given to laying out transistors and diodes in printed circuit boards. **Transistor leads** look different when viewed from the bottom (view *a*) than when viewed from the top (view *b*), as shown in Fig. 1.

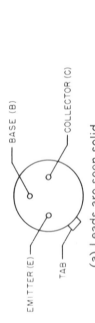

(a) Leads are seen solid.

(b) Leads are seen dotted.

FIG. I

The secret lies in the **tab** of the transistor. The emitter (E) is closest to the tab of the transistor. The **diode** is easier to lay out. Always observe in the schematic to which side the cathode ◄ is pointed. In the layout the diode should be installed in its proper orientation ⊢⊣⊢ or ⊢◄⊣. *It is important that the polarity of the diode be shown.*

Exercise: From the schematic shown below make a layout of the components as seen from the top of the board (the circuit will be dotted) and as seen from the circuit side (the components will be seen dotted). Show all missing components and lines. Show **transistor tabs** in bottom view. Lay out one component lead per land. Component outline will be found in Appendix B.

SCHEMATIC DIAGRAM

R2 47K 1/4W

CR2 1N617

Q2 2N1304

CR1 1N617

Q1 2N1304

R1 47K 1/4W

TOP VIEW
(component side)

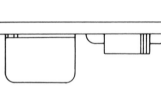

Q2

Q1

B
E
C

CR1

SIDE VIEW

BOTTOM VIEW
(circuit side)

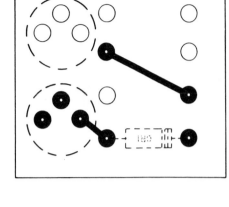

CRI

TITLE PRINTED CIRCUIT BOARD—TRANSISTORS AND DIODES

NAME | DATE | COURSE | GRADE | SCALE 2 X SIZE | DWG. NO. PCB-1 | PAGE 37

SHEET 2 OF 2

A TYPICAL, SINGLE-SIDED PCB (Printed circuit board)

STEP-BY-STEP FABRICATION

Printed circuit boards are fabricated in five principal steps, requiring five types of drawings:

1. **Schematic** 2. **Layout** 3. **Master** (Artwork) 4. **Drill & Contour** and 5. **Assembly** with **List of Materials**.

From specifications on a Schematic as shown at the upper right, a Layout started at the lower right can be drawn using components outlines similar to Appendix B.

Given: Schematic at right with component outline and Appendix B.

Problem: Prepare the drawings required to fabricate a PCB. Mount all components on a single-sided board 1.0 x 1.0 x .062 thick, and provide three PAN HEAD (#2-56 thread) SCREWS for installation.

Solution: Steps 1 and 2 are described at right while steps 3 through 5 are developed on page 39 & 40.

HOW TO PREPARE THE LAYOUT

Lay out your board and components double size for photographing. Remember that the transistor is viewed from top (see previous exercise). *In dotted line or red pencil* show the conductors between lands. Show the screw head, three places. Mark components and polarity of diodes as shown. *Always make a neat, accurate layout.*

The layout should show screws (as shown in PCB-2) or any other fastener which will occupy space on the P.C. Board and displace circuitry or component area. Although the screws have to be shown in the layout, they are not considered part of the assembly and should not appear in the List Of Materials.

SCHEMATIC OF PCB-2

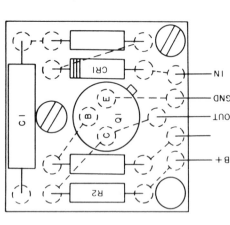

NOTE:
RESISTORS = 1/4 W, 5%
CAPACITOR = mmf, 500V

Step 1: SCHEMATIC (PCB-2)
Always draw the schematic first. Since this is a Sample Exercise, the schematic is shown completed.

Step 2: LAYOUT EXERCISE (PCB-2)
Complete the Layout.
1. How many electrical components are there? _____

LAYOUT OF PCB-2

2. Two components are not identified. Show them in the layout. (See schematic diagram.)
3. Show polarity of all diodes.
4. One screw head is not shown finished. Complete it.
5. Identify all leads coming out of the board.
6. Check your layout against the schematic and fill in the missing dotted line.

TITLE	PRINTED CIRCUIT BOARD–SCHEMATIC AND LAYOUT (SINGLE SIDED)		DWG. NO.	PCB-2	
NAME					PAGE
	DATE	COURSE	GRADE	SCALE 2 X SIZE	SHEET 1 OF 3	38

Step 4: PRINTED CIRCUIT BOARD DRILLED (PCB-2)

This step is needed so that the printed circuit board can be fabricated after the etching is done. Include the following data in all exercises to come:

1. Show all hole-diameter callouts.
2. Show board mounting hole center dimensions.
3. Show the shape and size of finished board dimensions.
4. **Material:** .062 thick Epoxy glass cloth laminate with .0027 CU one side (CU = copper)
5. **Finish:** Gold flash on etched lettering and conductors.

Exercise: See illustration below.

1. What should the .XX dimension be? _____
2. What are the A holes for? _____
   How many A holes are there? _____
3. What are the B holes for? _____
   How many B holes are there? _____
4. Complete the missing dimensions.
5. Show the two missing conductors as in step 3.

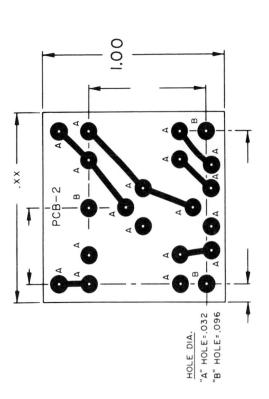

HOLE DIA.
"A" HOLE = .032
"B" HOLE = .096

Step 3: PCB MASTER (Artwork)

Generally, the layout made in Step 2 is turned over and a sheet of mylar is placed on the reverse side. The layout is usually made on semitransparent vellum whose PCB pattern is visible from the back side through the mylar. The PCB pattern is now duplicated on the mylar with black tape conductors and/or decal lands. If you had made a master of PCB-2, it would be similar to the sample shown below.

The master (artwork) should show the following:

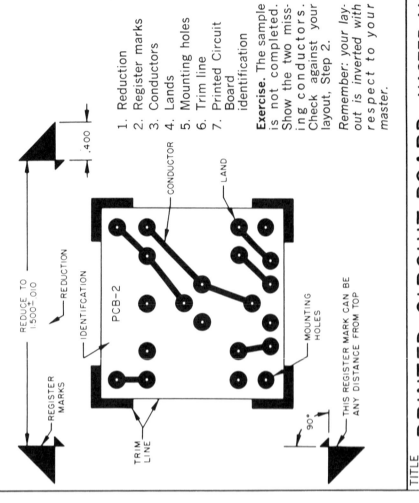

REDUCE TO
1.500 ± .010 — REDUCTION

.400

IDENTIFICATION

PCB-2

CONDUCTOR

LAND

REGISTER MARKS

TRIM LINE

MOUNTING HOLES

90°

THIS REGISTER MARK CAN BE ANY DISTANCE FROM TOP

1. Reduction
2. Register marks
3. Conductors
4. Lands
5. Mounting holes
6. Trim line
7. Printed Circuit Board identification

Exercise. The sample is not completed. Show the two missing conductors. Check against your layout, Step 2.

Remember: your layout is inverted with respect to your master.

Step 5: PRINTED CIRCUIT BOARD ASSEMBLY (PCB-2)

The following items must be shown:

1. Assembly of all components as seen mounted on board. It will be the same as your layout, except that the dotted line and lands will not be seen since it is on the opposite side of the board.

2. All components and identification. *Always make it clear so that anyone could assemble the board from your assembly drawings.*

3. The assembly drawing should always be accompanied by a complete list of material, as shown below.

Exercise

1. Complete the assembly drawing below. Show all component identifications.

2. Complete the list of material (see Appendix B and schematic diagram on page 38).
   Resistors are listed on page 88 (MIL STYLE RC07).

   Capacitors are listed on page 85 (MIL TYPE CM-15).

   Diodes are listed on page 87.

   Transistors are listed on page 91.

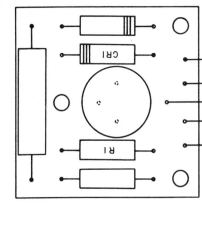

ITEM	NO. REQ.D	REFERENCE DESIGNATION	DESCRIPTION	MANUFACTURER & PART NO. OR MIL TYPE DESIGNATION
7		R2	RESISTOR	6.8K, ¼W, 5% RC07GF682J
6				
5				
4	2	CR1, CR2	DIODE	
3	1	C1	CAPACITOR	50 MMF, 500V, CM
2	1	PCB-2	PRINTED BOARD	1.00 x 1.00 x .062 EPOXY GLASS
1	X	– – – – –	SCHEMATIC	SHEET 1 of 3

LIST OF MATERIAL

TITLE	PRINTED CIRCUIT BOARD – ASS'Y AND LIST OF MATERIAL (SINGLE SIDED)	DWG. NO. **PCB-2**		PAGE
NAME				**40**
DATE	COURSE	GRADE	SCALE	SHEET **3** OF **3**
			2 X SIZE	

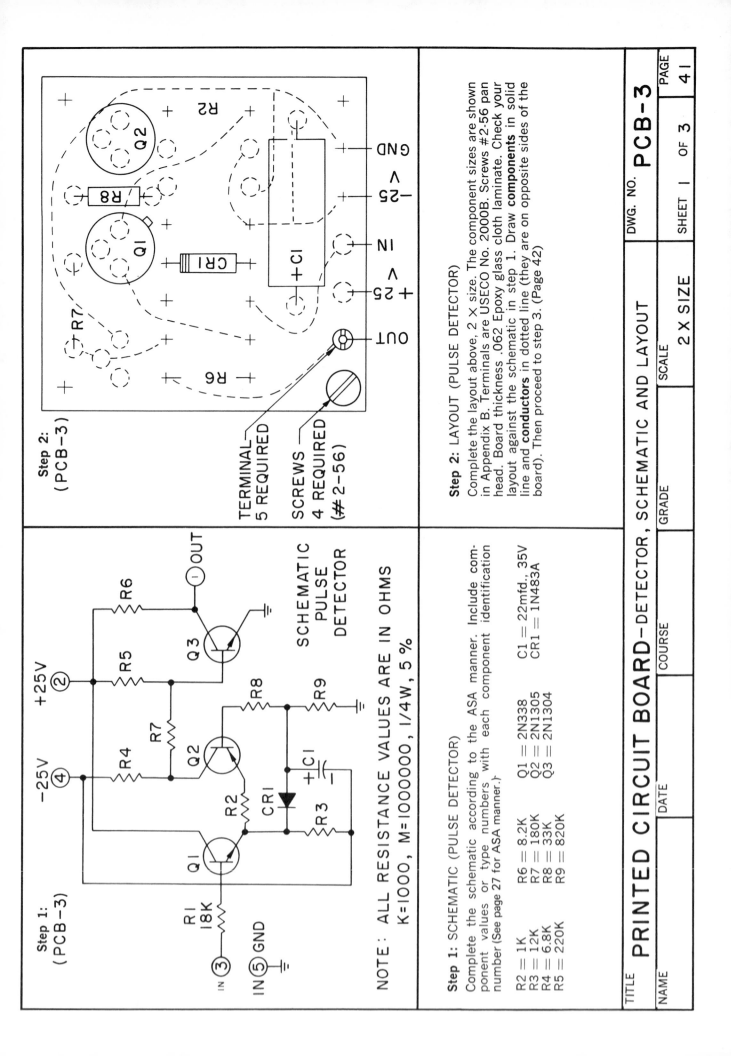

Step 2:
(PCB-3)

R2
Q2
R8
Q1
CR1
R7
R6
+C1

GND
−25 V
IN
+25 V
OUT

TERMINAL
5 REQUIRED

SCREWS
4 REQUIRED
(#2-56)

Step 2: LAYOUT (PULSE DETECTOR)

Complete the layout above, 2 X size. The component sizes are shown in Appendix B. Terminals are USECO No. 2000B. Screws #2-56 pan head. Board thickness .062 Epoxy glass cloth laminate. Check your layout against the schematic in step 1. Draw **components** in solid line and **conductors** in dotted line (they are on opposite sides of the board). Then proceed to step 3. (Page 42)

Step 1:
(PCB-3)

+25 V ②
−25 V ④
R6
R5
R4
R7
R2
R8
R9
CR1
+C1
R3
R1 18K
Q1
Q2
Q3
① OUT
IN ⑤ GND
③ IN

SCHEMATIC
PULSE
DETECTOR

NOTE: ALL RESISTANCE VALUES ARE IN OHMS
K=1000, M=1000000, 1/4W, 5%

Step 1: SCHEMATIC (PULSE DETECTOR)

Complete the schematic according to the ASA manner. Include component values or type numbers with each component identification number (See page 27 for ASA manner.)

R2 = 1K	R6 = 8.2K	Q1 = 2N338	C1 = 22mfd., 35V
R3 = 12K	R7 = 180K	Q2 = 2N1305	CR1 = 1N483A
R4 = 6.8K	R8 = 33K	Q3 = 2N1304	
R5 = 220K	R9 = 820K		

TITLE PRINTED CIRCUIT BOARD-DETECTOR, SCHEMATIC AND LAYOUT

DWG. NO. PCB-3

PAGE 41

NAME

DATE

COURSE

GRADE

SCALE 2 X SIZE

SHEET 1 OF 3

Step 3: PRINTED CIRCUIT BOARD MASTER (ARTWORK)

Complete the master, following the same method used in sample exercise, PCB-2. Show reduction. (Reduce to 2.000 ±.010, which will reduce the board to half size giving a 1.750 x 1.750 full-size board.) Show register marks, conductors, etc.

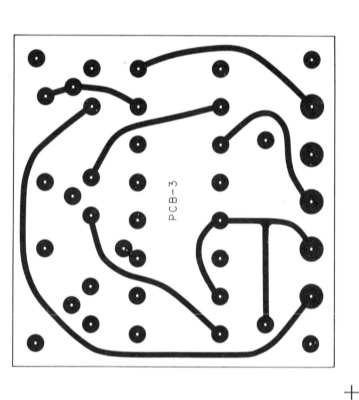

PCB-3

Step 4: PRINTED CIRCUIT BOARD DRILLED

Follow the same method used in sample exercise PCB-2. Complete the drawing of the board below. For clarity do not show conductors again but show all dimensions necessary for fabrication of board.

Hole diameter

A (land hole) = .032 Dia.
C (terminal mounting holes) = .065 $^{+.003}_{-.001}$ Dia.

Calculate **clearance B holes,** four places for #2-56 pan head screws (Use Formula, Appendix D – use "B" ℄ – ℄ tol. = ±.005).

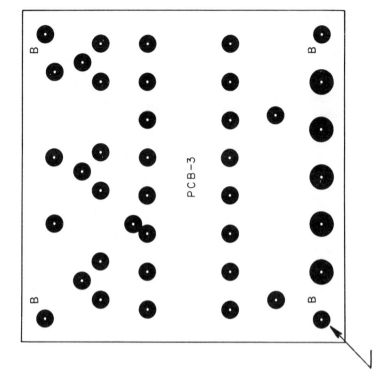

PCB-3

TITLE	PRINTED CIRCUIT BOARD—DETECTOR, MASTER AND DRILLED BOARD		DWG. NO. PCB-3	
NAME	DATE	COURSE	GRADE	SCALE 2 X SIZE	SHEET 2 OF 3	PAGE 42

PRINTED CIRCUIT BOARD
ASSEMBLY
AND LIST OF MATERIAL ②

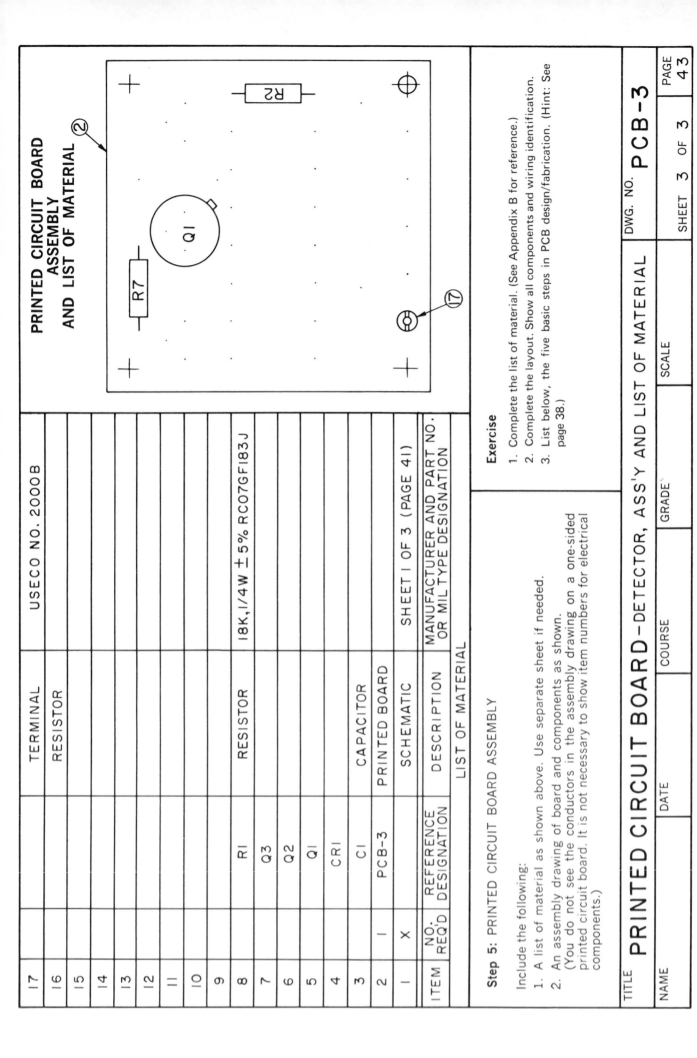

ITEM	NO. REQ'D	REFERENCE DESIGNATION	DESCRIPTION	MANUFACTURER AND PART NO. OR MIL TYPE DESIGNATION
17			TERMINAL	USECO NO. 2000 B
16			RESISTOR	
15				
14				
13				
12				
11				
10				
9				
8		R1	RESISTOR	18K,1/4W ±5% RC07GF183J
7		Q3		
6		Q2		
5		Q1		
4		CR1		
3		C1	CAPACITOR	
2	1	PCB-3	PRINTED BOARD	
1	X		SCHEMATIC	SHEET 1 OF 3 (PAGE 41)

LIST OF MATERIAL

Step 5: PRINTED CIRCUIT BOARD ASSEMBLY

Include the following:

1. A list of material as shown above. Use separate sheet if needed.
2. An assembly drawing of board and components as shown.
   (You do not see the conductors in the assembly drawing on a one-sided printed circuit board. It is not necessary to show item numbers for electrical components.)

Exercise

1. Complete the list of material. (See Appendix B for reference.)
2. Complete the layout. Show all components and wiring identification.
3. List below, the five basic steps in PCB design/fabrication. (Hint: See page 38.)

TITLE	PRINTED CIRCUIT BOARD–DETECTOR, ASS'Y AND LIST OF MATERIAL			DWG. NO. PCB-3	
NAME		DATE	COURSE	GRADE	SCALE

SHEET 3 OF 3

DOUBLE-SIDED PRINTED CIRCUIT BOARDS

Double-sided PCBs have the following features:

1. They are more expensive to manufacture than single-sided PCBs.

2. They are used where space is limited.

3. In double-sided boards the components are on one or both sides of the board.

4. Conductors are on both sides of the board.

5. An accurate layout is required in order to line up the front and back faces of the board; therefore, tool holes are required for machining.

CONNECTING THE FRONT AND BACK FACE OF THE BOARD

The four general methods for accomplishing connection between two sides of a double-sided PCB are (1) leads, (2) eyelets, (3) plated-through holes, and (4) terminals.

Exercise (Step 1 and Step 2 PCB-4)

From the same schematic diagram Pulse Detector and list of material in Exercise PCB-3, Fabricate a double-sided PCB.

Given:

1. **Schematic**, Exercise PCB-3, Step 1. (Page 41)

2. **List of material**, Exercise PCB-3, Step 5. (Page 43)

3. **PCB-4**, 1.750 x 1.250 x .062 thick. (Notice the board is now smaller in size than PCB-3.)

PROBLEM: Complete the started layout of PCB-4 below. Show transistor tabs in proper location. **Three components**, shown with hidden lines, are mounted on the back side of the board.

FRONT FACE : SOLID LINES ————
BACK FACE : HIDDEN LINES — — —

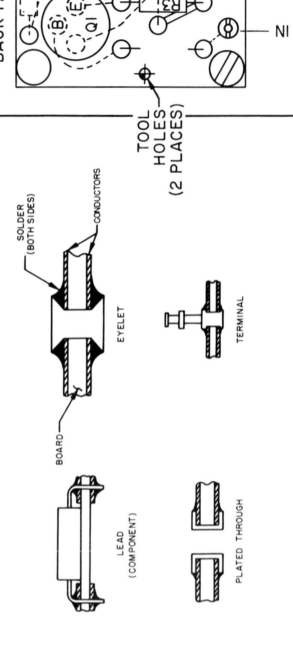

TOOL HOLES (2 PLACES)

SOLDER (BOTH SIDES)

CONDUCTORS

EYELET

BOARD

TERMINAL

LEAD (COMPONENT)

PLATED THROUGH

Step 3: MASTER (ARTWORK) PCB-4

The method is the same as in the single-sided board, except that in the double-sided PCB **two masters** (two pieces of artwork) are needed for photographing.

Exercise: PCB-4

Complete the two masters (both sides of PCB-4) below. Follow the same method used in preparing the master (artwork) of PCB-3 (step 3).

The **front-face master** is prepared from the layout with *solid lines*.

The **back-face master** is prepared from the layout with *hidden lines*.

Show the correct reduction of PCB-4 for both sides (**Hint:** board size is 1.750 x 1.250 x .062) Tool holes are already shown in both masters.

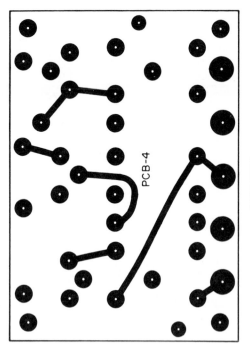

PCB-4

BACK FACE

FRONT FACE

TITLE PRINTED CIRCUIT BOARD–DOUBLE SIDED, BOARD MASTERS

DWG. NO. PCB-4

PAGE 45

SHEET 2 OF 4

SCALE 2 X SIZE

NAME | DATE | COURSE | GRADE

Step 4: PCB DRILLED (DOUBLE-SIDED PCB-4)

The method is the same as in a single-sided board. Choose the master with the identification PCB-4. Since the lands on both sides of the PCB are aligned, drilling may be done through either side.

Exercise: PCB-4

Complete the PCB-4 drilled drawing below in the same manner as was used in Exercise PCB-2 (step 4). Add the tool holes (.062 Dia.) and complete all the notes, hole chart, etc.

(Hint: See page 39 for Material & Finish)

NOTES

MATERIAL : .062 THICK EPOXY _____

FINISH : GOLD _____

_____ BOTH SIDES.

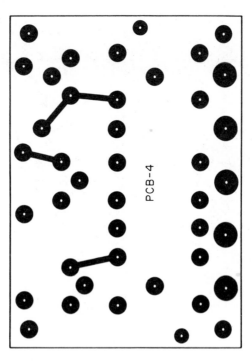

PCB-4

HOLE CHART

DESCRIPTION	LETTER	DIA.	NO. REQ'D
LAND HOLES	A		
BOARD MTG. HOLE	B		
TERMINAL HOLE	C	$.065^{+.003}_{-.001}$	
TOOL HOLES	D		2

TITLE PRINTED CIRCUIT BOARD – DOUBLE SIDED, BOARD DRILLED

DWG. NO. PCB-4

PAGE 46

NAME	DATE	COURSE	GRADE	SCALE 2 X SIZE

Step 5: ASSEMBLY (DOUBLE-SIDED PCB-4)

The assembly is prepared in the same manner as the single-sided board PCB-3 (step 5). The only difference is that in the double-sided PCB two sides are to be shown — the front and the back — with all the component identifications. Since this exercise has the same list of material as in PCB-3, it is not necessary to show it again.

Exercise: PCB-4

Complete the assembly drawing below by showing all the components. Some of the conductors have been omitted for clarity. Do not draw them in.

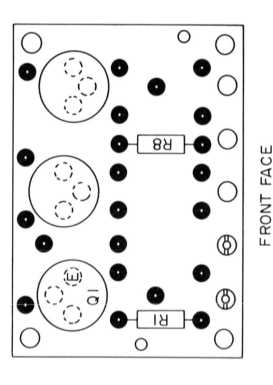

BACK FACE

FRONT FACE

TITLE	PRINTED CIRCUIT BOARD—DOUBLE SIDED, BOARD ASSEMBLY			DWG. NO. **PCB-4**	
NAME	DATE	COURSE	GRADE	SCALE 2 X SIZE	PAGE 47
					SHEET 4 OF 4

Exercise: PCB-5 (PRINTED CIRCUIT BOARD)

From the schematic of the **Flip-Flop (PCB-5)** shown below, prepare all the drawings required (four more) to fabricate a **Single-Sided PCB-5** (see exercises PCB-3).

Exercise: PCB-5

Start with the Layout below (2 × size). Make the length of the board layout as compact as possible. Spacing between components, land, conductors, etc. should be ⅟₃₂ minimum (⅟₁₆ in 2 × size layout). All required data may be found in Appendixes (outline and hardware) B and C. Identify this board as PCB-5.

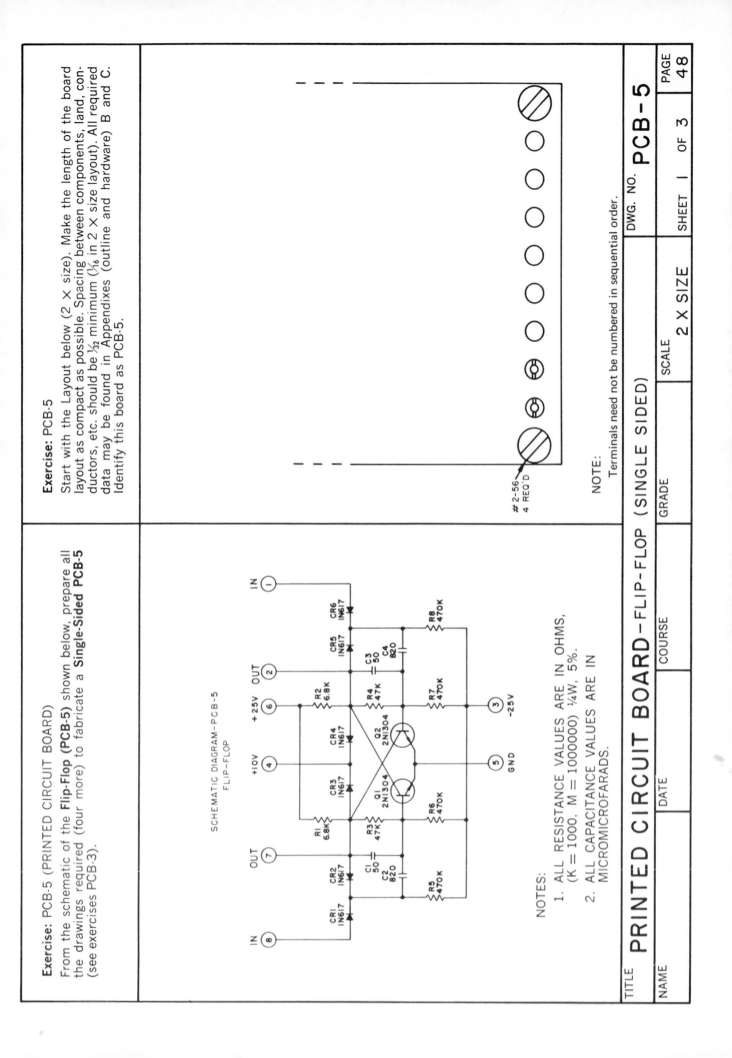

SCHEMATIC DIAGRAM-PCB-5
FLIP-FLOP

NOTES:
1. ALL RESISTANCE VALUES ARE IN OHMS, (K = 1000, M = 1000000) ¼W, 5%.
2. ALL CAPACITANCE VALUES ARE IN MICROMICROFARADS.

2-56
4 REQ'D

NOTE:
Terminals need not be numbered in sequential order.

TITLE					
PRINTED CIRCUIT BOARD–FLIP-FLOP (SINGLE SIDED)				DWG. NO. **PCB-5**	
NAME	DATE	COURSE	GRADE	SCALE	PAGE
				2 X SIZE	48
				SHEET 1 OF 3	

Exercise: PCB-5 (MASTER, ARTWORK)

Complete the master of PCB-5 below. Use either pencil or tapes, as instructed by your teacher.

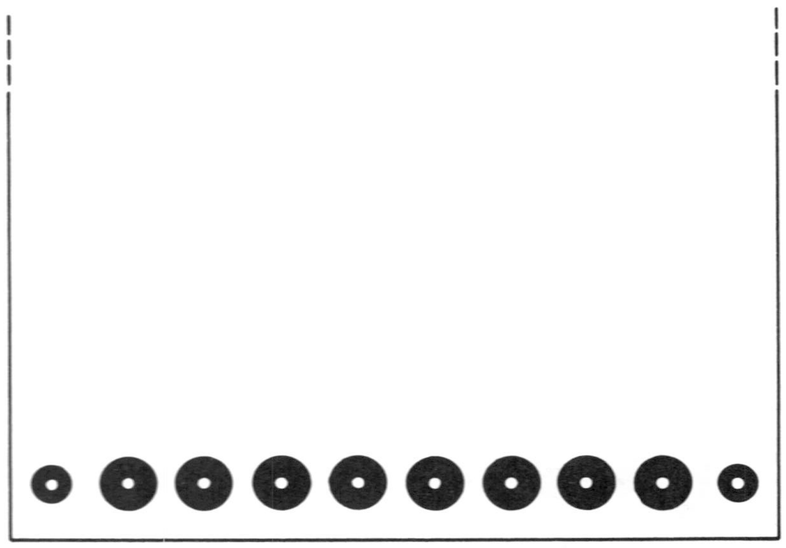

Exercise: PCB-5 (DRILLED)

Complete the drilled board of PCB-5 below.

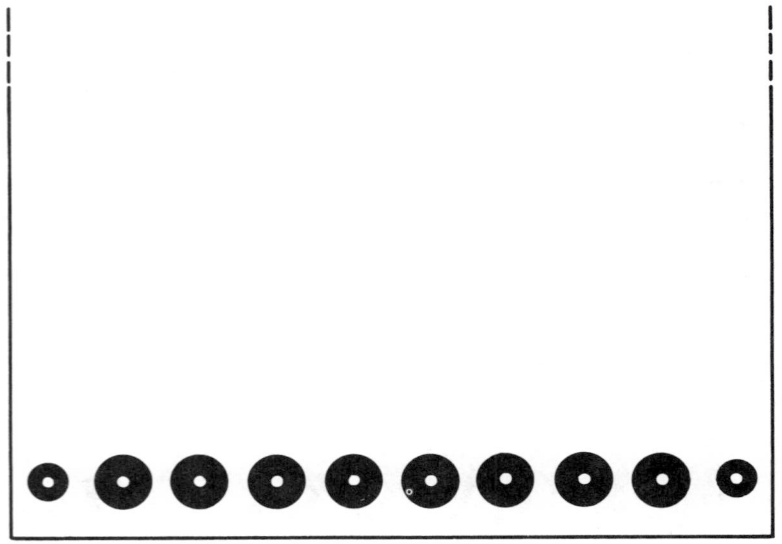

TITLE PRINTED CIRCUIT BOARD—FLIP-FLOP (SINGLE SIDED)

DWG. NO. PCB-5

PAGE 49

NAME

DATE

COURSE

GRADE

SCALE 2 X SIZE

SHEET 2 OF 3

LIST OF MATERIAL

ITEM	NO. REQD.	REFERENCE DESIGNATION	DESCRIPTION	MANUFACTURER AND PART NUMBER OR MIL — TYPE DESIGNATION
1	X		SCHEMATIC	SHEET 1 OF 3
2		PCB-5		
3		C1, C3		
8				
		R5, R6, R7, R8		

Exercise: PCB-5 (ASSEMBLY)

Complete the assembly and list of material of PCB-5 below.

TITLE PRINTED CIRCUIT BOARD — FLIP-FLOP (SINGLE SIDED)

DWG. NO. PCB-5

NAME	DATE	COURSE	GRADE	SCALE 2 X SIZE	SHEET 3 OF 3	PAGE 50

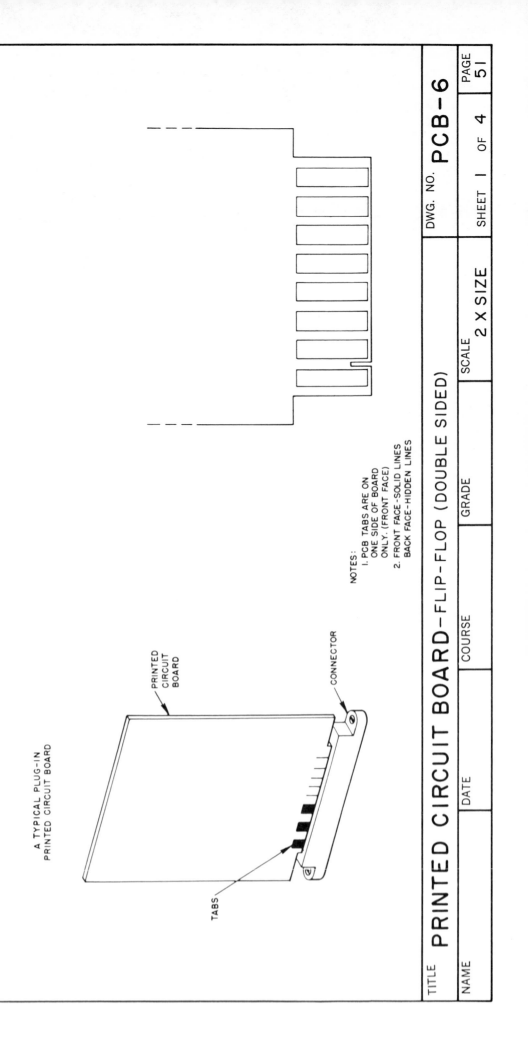

PRINTED CIRCUIT BOARD PCB-6.

From the preceding schematic and the same list of material of PCB-5, prepare all the drawings required (four or more) to fabricate a **double-sided PCB-6** (see PCB-4).

Exercise: (PCB-6)

Start with the layout below (2 × size). Make the length of the board layout as compact as possible, smaller than PCB-5. Use tabs in this fabrication so that the board can be plugged into a connector. No mounting screws are required. Spacing between components, land, conductors, etc., should be $\frac{1}{32}$ minimum ($\frac{1}{16}$ in 2 × size layout). Use the same data as in the preceding exercise (PCB-5). Identify this board as PCB-6.

A TYPICAL PLUG-IN
PRINTED CIRCUIT BOARD

PRINTED
CIRCUIT
BOARD

TABS

CONNECTOR

NOTES::
1. PCB TABS ARE ON
   ONE SIDE OF BOARD
   ONLY. (FRONT FACE)
2. FRONT FACE–SOLID LINES
   BACK FACE–HIDDEN LINES

TITLE	PRINTED CIRCUIT BOARD–FLIP-FLOP (DOUBLE SIDED)				DWG. NO. **PCB-6**	
NAME	DATE	COURSE	GRADE	SCALE 2 X SIZE	SHEET 1 OF 4	PAGE 51

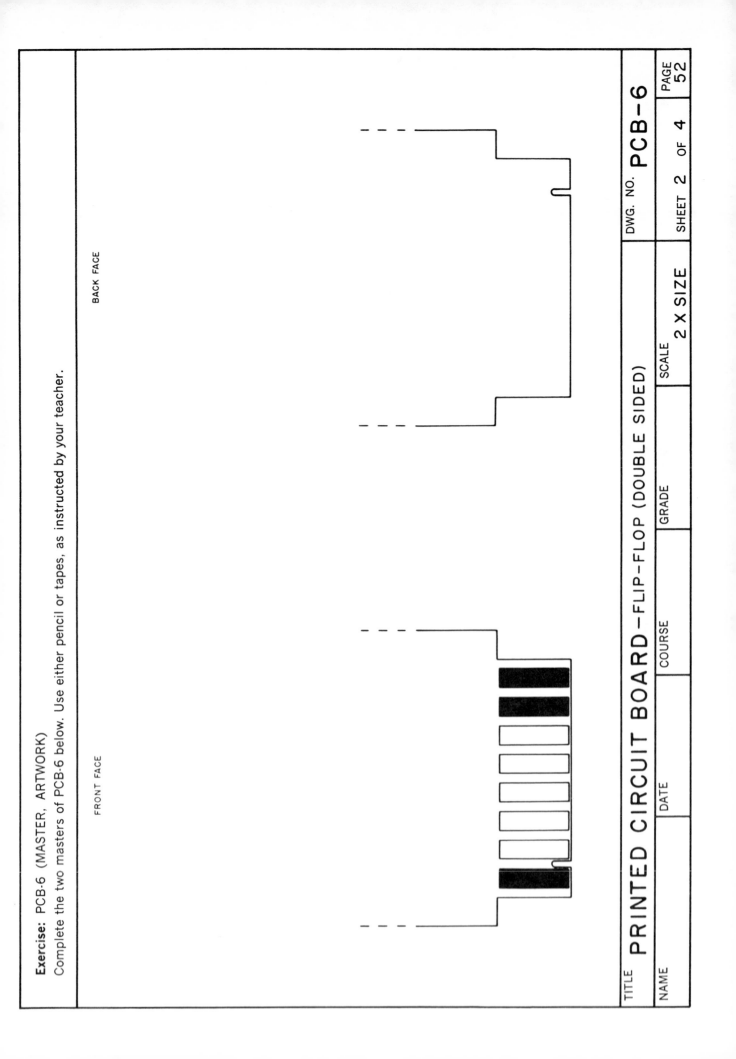

Exercise: PCB-6 (MASTER, ARTWORK)

Complete the two masters of PCB-6 below. Use either pencil or tapes, as instructed by your teacher.

FRONT FACE

BACK FACE

TITLE PRINTED CIRCUIT BOARD—FLIP—FLOP (DOUBLE SIDED)

DWG. NO. **PCB-6**

NAME	DATE	COURSE	GRADE	SCALE	
				2 X SIZE	SHEET 2 OF 4

PAGE 52

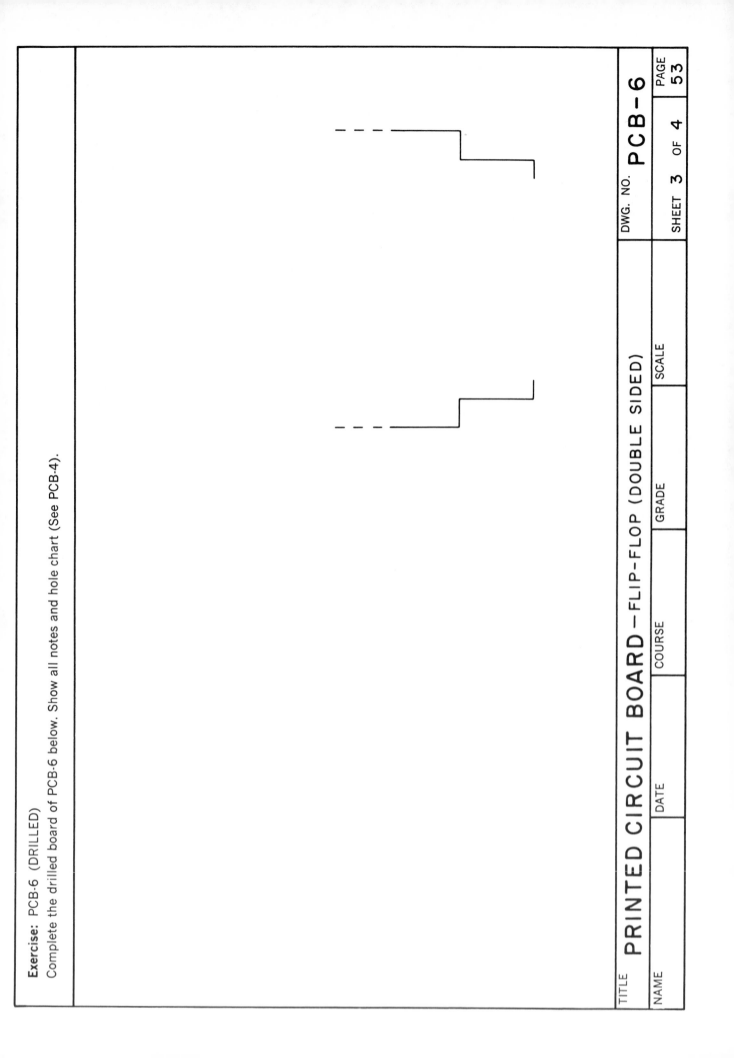

Exercise: PCB-6 (DRILLED)

Complete the drilled board of PCB-6 below. Show all notes and hole chart (See PCB-4).

TITLE PRINTED CIRCUIT BOARD – FLIP–FLOP (DOUBLE SIDED)

DWG. NO. **PCB-6**

NAME	DATE	COURSE	GRADE	SCALE

SHEET **3** OF **4**

PAGE **53**

Exercise: PCB-6 (ASSEMBLY)

Complete the assembly of both sides (front and back) of PCB-6 below. Since this exercise has the same list of materials as PCB-5, it is not necessary to show it again. Draw only the components and outline of boards; do not show lands or conductors.

BACK FACE

FRONT FACE

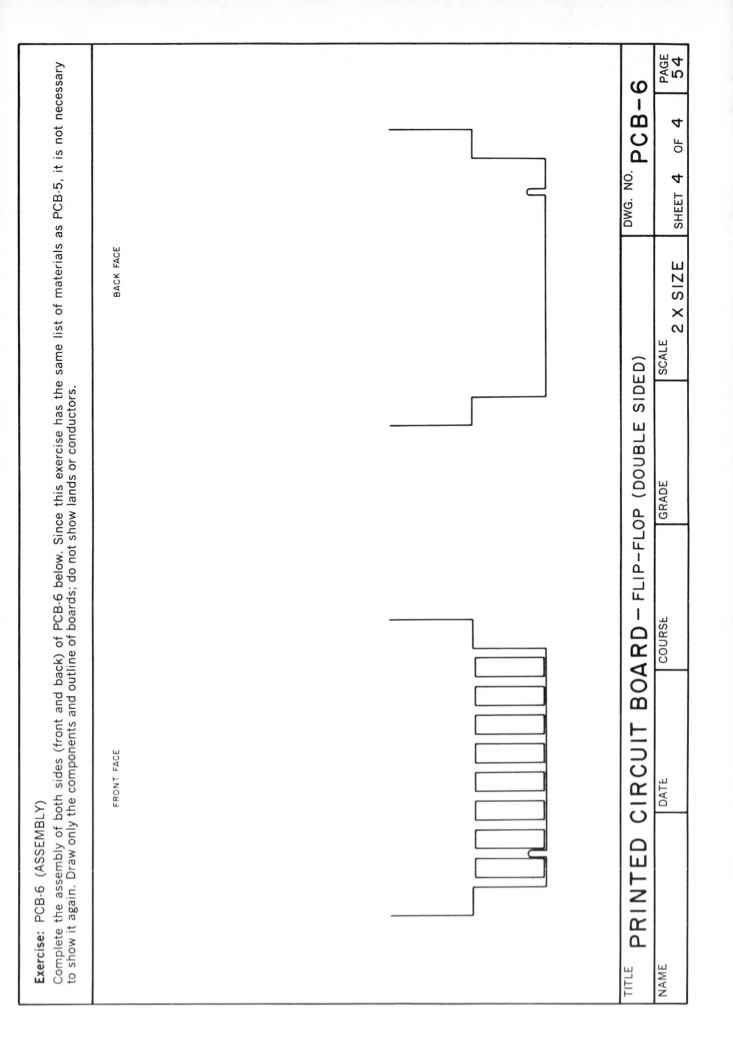

TITLE	PRINTED CIRCUIT BOARD – FLIP-FLOP (DOUBLE SIDED)		DWG. NO. PCB-6	
NAME	DATE	COURSE	GRADE	SCALE 2 X SIZE	SHEET 4 OF 4	PAGE 54

INTRODUCTION: ELECTROMECHANICAL DESIGN

Basic steps needed in Electromechanical design are:

1. **Schematic diagram** (given to you in sketch form)
2. **Preliminary electrical parts list** (given)
3. **Layout** (done by you as designer)
4. **Detail drawings** (mechanical, from catalogues and/or direct measurements)
5. **Final assembly drawing** and **complete list of material**
6. **Wiring diagram** (both steps 5 and 6 done by you)

POWER SUPPLY DESIGN

Exercise. An electronics engineer wants you to design a simple package of the power supply given in the schematic shown in Fig. 1.

Solution. The first thing to do is to study the schematic. Then make a **prelim-inary electrical parts list** and **outline** dimensions of all the **components** given in the schematic.

Complete the preliminary electrical parts list by selecting the proper electrical components from Appendix B.

Next, devise a simple, neat, easy-to-build, serviceable package (continued on the next pages).

Note: XF1 is related to F1 by the following MIL-STD definition: A socket, fuseholder, or similar device that is always associated with a single particular part or subassembly (such as an electron tube, a fuse etc.) shall be identified by a composite reference designation consisting of the class letter "X," followed by the basic reference designation that identifies the mounted part. For example, the socket for fuse F1 would be XF1.

POWER SUPPLY SCHEMATIC DIAGRAM

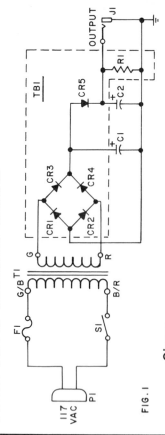

FIG. 1

Given:

Capacitor, C1 & C2 = 1500mfd, 50V
Diode, CR1 Thru CR5 = 1N91
Resistor, R1 = 22K, ½W, 5%
Transformer, T1 = TRIAD No. F-92A
Fuse, F1 = 1 amp, 125V (3 AG)
Fuse Holder, XF1 = Littelfuse, No. 342001
Toggle switch, S1 = MS35058-22
Output Jack, J1 = Pin Jack Strip (two pins)
Power cord, P1 = any standard cord (black)

PRELIMINARY ELECTRICAL PARTS LIST AND OUTLINE DIMENSIONS OF COMPONENTS

See Appendix B and complete the list below:

C1 and C2 = 1⅛ dia. × 3⁹⁄₁₆ high (ARCO)

CR1 through CR5 = _____

P1 = ¼ O.D. cord _____ 6 FEET LONG, STOCK ITEM

R1 = _____ (RC20GF223J)

T1 = _____

XF1 = _____

F1 = _____ (LITTELFUSE, INC.)

S1 = _____

J1 = _____

ELECTROMECHANICAL DESIGN (cont.)

From your preliminary **electrical parts list** you will notice that transformer (T1) and capacitors (C1 and C2) are the largest components. One must select an enclosure or chassis large enough to house all the components and necessary hardware. This brings us to the third step — namely, the **layout.**

LAYOUT

The layout should describe the parts sufficiently for the detailer or a draftsman to understand what has been selected. Common screws or any other hardware may be written on the layout near the part for clarity.

The layout should be drawn accurately to scale. In this case we will use half scale. Other accessories which will be needed in our package will include the following:

*1. **Terminal board** (TB1) for mounting all diodes and resistor.

2. **Fuse holder** to house fuse (F1).

3. **Stand-offs, grommet, terminal lug, screws, nuts, washers,** etc.

You will find that a **chassis** $5\frac{1}{2} \times 4\frac{3}{4} \times 1\frac{9}{16}$ would be large enough for our purpose.

Exercise. Complete the two views of the started layout on the right (scale: half size). Identify all components with reference designations. See Schematic. (Page 55)

In general, additional information required for the design and/or details (such as **material for the chassis, finishes, specs,** etc.) are given in note form.

*Make terminal board (TB1) 3.50 inches long with CR1, CR2, CR3, and CR4 mounted in the same manner as R1 and CR5.

CHASSIS LAYOUT.

GROMMET

CHASSIS

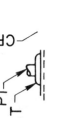

TOP VIEW

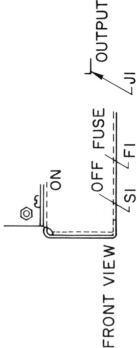

SOLDER
LUG NO. 8

NOTES:

1. Material: alum sheet, $\frac{1}{16}$ thk. 6061-T6, per QQ-A-327, cond T.

2. Finish: alodine per MIL-C-5541 external surfaces and edges shall be painted gray.

3. Silk screen or rubber stamp letters on front, black.

4. No. 8-32 screws are used on T1 and J1 (4 and 2 req'd). No. 4-40 screws and stand-offs are used on TB1 (4 req'd).

FRONT VIEW

TITLE	ELECTROMECHANICAL DESIGN–LAYOUT			
NAME	DATE	COURSE	GRADE	PAGE
				56
	DWG. NO. **EMD-1**			
	SCALE 1/2	SHEET 2 OF 2		

ELECTROMECHANICAL DESIGN (cont.)

From the layout we proceed to the fourth step — namely, the **details.** The details are drawn by a draftsman or detailer from the layout. Detail parts should show dimensions, tolerance, etc. Whenever possible, for clarity, show the detail as large as space and standard scaling permit (e.g., 2/1 or 4/1). Use as many views and sections as are necessary to completely describe the part. It would be a good idea to detail the **terminal board** (TB1) before you detail the chassis.

Exercise: TERMINAL BOARD (TB1) DETAIL

From the layout you will notice that six components are mounted on the terminal board: five **diodes** and one **resistor.** That means that twelve **terminals** are required (two terminals per component).

Draw the terminal board full scale. Terminals are USECO No. 2000 B. (See Hardware, Appendix C, page 96). Show the board, terminals, and markings all in one detail.

Board dimensions are 1.40 × 3.50 × .062 thick.

Terminal spacings are .50 and 1.00.

Dimension the terminal board according to standard practice as in Lesson 1, (Mechanical Drafting Review). Silk Screen or rubber stamp terminals from 1 through 12 as shown. Identify board TB1 parallel to right edge. In practical applications the drilling and marking details are done separately, but both details are combined here for simplicity.

After completing the details, proceed to the next page for **terminal board assembly.**

Exercise. Complete the two views of the terminal board below.

NOTES:
1. TERMINAL BOARD MATERIAL: .062 THK. PHENOLIC
2. SILK SCREEN OR RUBBER STAMP NUMBERS AND TERMINAL BOARD IDENTIFICATION (TB1) ⅛ HIGH, BLACK, BEFORE INSTALLING TERMINALS.
3. INSERT TERMINALS INTO BOARD AND SWAGE THEM OVER.

TITLE	ELECTROMECHANICAL DESIGN–TERMINAL BOARD DETAIL		DWG. NO. EMD-2		
NAME	DATE	COURSE	GRADE	SCALE FULL	PAGE 57
					SHEET 1 OF 2

TERMINAL BOARD ASSEMBLY

The **assembly board** is drawn in the same manner as has been done in previous lessons. The only difference is that the Assembly here is drawn in two stages. This is sometimes called a **stage drawing.**

Stage 1. In this step, only the **board** and the **jumper wires** are shown. The terminals are connected with jumper wires. You will have to check this against the schematic and layout to determine which terminal will require the jumpers. Also show wires coming out of the board and identify the routes of these wires by assigning a **color code** to each. The color code will later help the technician to complete the wiring of the **power supply.** There are three wires coming out of the board:

From junction CR1, CR2, and R1, one wire (black)
From junction CR3, CR4, and CR5, one wire (red)
From junction R1 and CR5, one wire (green)

These wires should be at least 10″ long.

Stage 2. Draw the board again, but in this view show only the **electrical components** (CR1 through CR5 and R1). Show reference designations of all components. They should conform to those on the schematic diagram. With the assembly, include a list of material.

Note: If this board had been made into a **printed circuit board,** stage 1 would have been omitted. Due to the simplicity of the board, wires were used.

Complete the exercises on the right.

Exercise. Complete two stages and the list of material below.

Stage 1

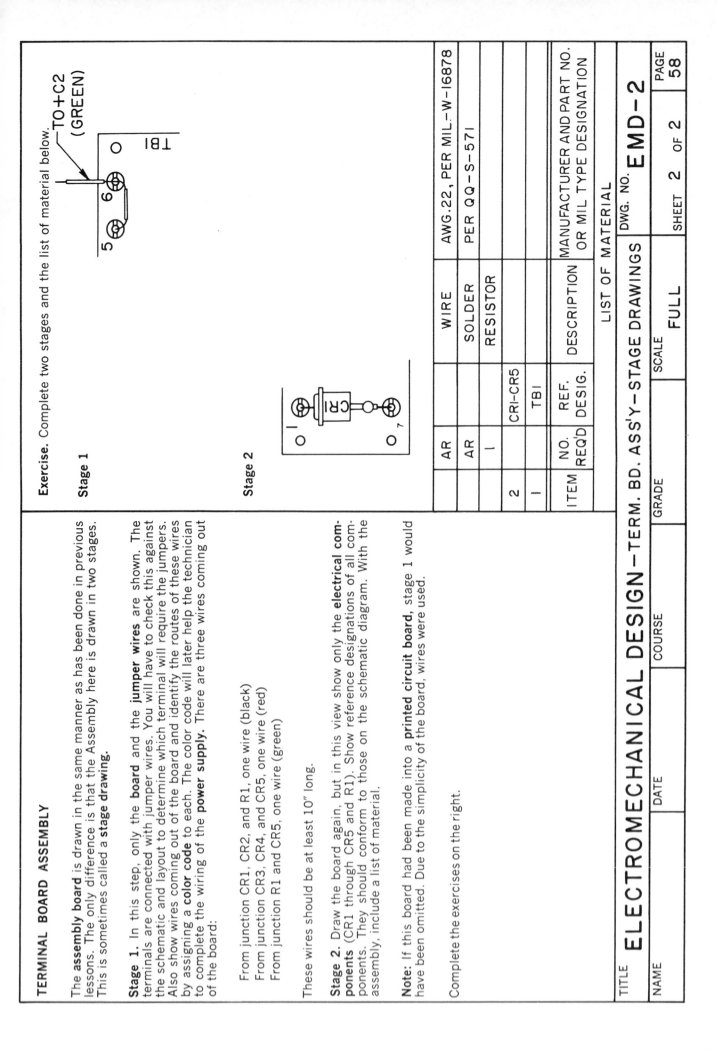

Stage 2

	NO. REQ'D	REF. DESIG.	DESCRIPTION	MANUFACTURER AND PART NO. OR MIL TYPE DESIGNATION
	AR		WIRE	AWG.22, PER MIL.-W-16878
	AR		SOLDER	PER QQ-S-571
	1		RESISTOR	
	2	CRI-CR5		
	1	TBI		
ITEM	NO. REQ'D	REF. DESIG.	DESCRIPTION	MANUFACTURER AND PART NO. OR MIL TYPE DESIGNATION

LIST OF MATERIAL

TITLE				DWG. NO.	
ELECTROMECHANICAL DESIGN—TERM. BD. ASS'Y—STAGE DRAWINGS				EMD-2	
NAME	DATE	COURSE	GRADE	SCALE FULL	SHEET 2 OF 2	PAGE 58

Exercise. Complete the drawing of the **chassis** below showing three views (front, top, and back) plus Detail H (hole pattern for C1 and C2), and fill in **hole chart**. (See Lesson #1 for reviewing hole-pattern details). Scale: half size

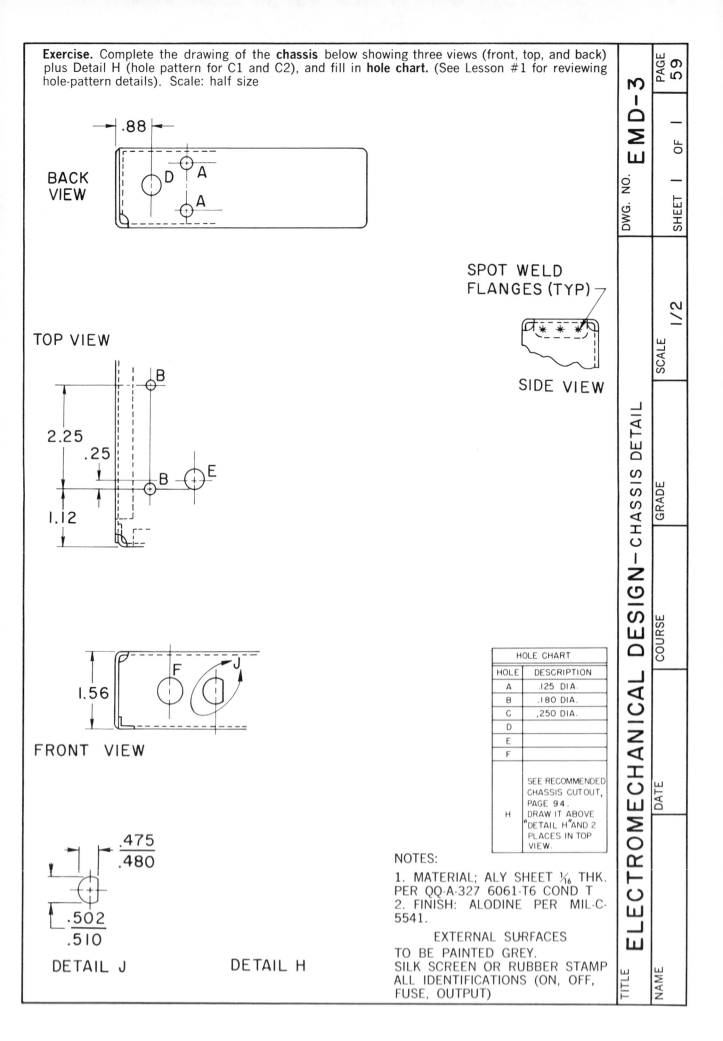

BACK VIEW

.88

D   A

A

TOP VIEW

2.25
.25
1.12
B
B   E

SPOT WELD FLANGES (TYP)

SIDE VIEW

1.56
F   J

FRONT VIEW

.475
.480
.502
.510

DETAIL J

DETAIL H

HOLE CHART	
HOLE	DESCRIPTION
A	.125 DIA.
B	.180 DIA.
C	.250 DIA.
D	
E	
F	
H	SEE RECOMMENDED CHASSIS CUTOUT, PAGE 94. DRAW IT ABOVE "DETAIL H" AND 2 PLACES IN TOP VIEW.

NOTES:

1. MATERIAL; ALY SHEET 1/16 THK. PER QQ-A-327 6061-T6 COND T
2. FINISH: ALODINE PER MIL-C-5541.

EXTERNAL SURFACES TO BE PAINTED GREY. SILK SCREEN OR RUBBER STAMP ALL IDENTIFICATIONS (ON, OFF, FUSE, OUTPUT)

ELECTROMECHANICAL DESIGN—CHASSIS DETAIL

GRADE   COURSE   DATE   TITLE   NAME

ASSEMBLY DRAWINGS

Assembly drawings should be drawn so that the detailed parts and **sub-assemblies** are shown in their relative position and scale with the number of views necessary to clearly portray the proper attachments. All detailed parts and sub-assemblies should be identified by an **item number** and, if an electrical part, by marking its **circuit designation** on the component. The item number should be placed in a $\frac{3}{8}$ diameter circle with a lead running from the circle to the part it identifies.

LIST OF MATERIAL

For each assembly drawing prepare a **list of material** and an **electrical parts list** containing all items which become part of the completed assembly, including (1), every item purchased or fabricated and (2), all material including finishes.

A subassembly which becomes part of the assembly should be listed as one item and identified by its drawing number (for example, the terminal board TB1). The list of material (hardware and fabricated parts) and the list of electrical parts should be prepared separately.

Exercise. Complete the **assembly drawing** on the right. If you made a good layout it could be used as the assembly drawing.

Show all items and identify the electrical components with **reference designations.** The reference designations of all electrical components should conform to those on the schematic diagram.

On the next page prepare separately, a list of material and an electrical parts list. For references use the schematic diagram, layout, and the appendixes.

The wiring diagram of the "power supply" will be drawn in Lesson 10 on wiring diagrams.

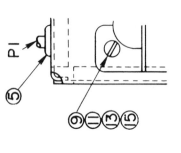

TITLE		ELECTROMECHANICAL DESIGN–ASSEMBLY DRAWING		DWG. NO.	EMD-4
NAME	DATE	COURSE	GRADE	SCALE 1/2	PAGE 60
					SHEET 1 OF 2

Complete the List of Material and the Electrical Parts List below. Use the assembly drawing and appendixes for reference.

ELECTRICAL PARTS LIST

ITEM	NO. REQ'D	REF. DESIG	DESCRIPTION	MANUFACTURER AND PART NO. OR MIL TYPE DESIGNATION
10	AR		HOOKUP WIRE	AWG. 22, PER MIL-W-16878
9	AR		SOLDER SOFT	PER QQ-S-571
8		T1		TRIAD TRANSFORMER CORPORATION NO. F-92A
7	1		SWITCH TOGGLE	
6	1		PLUG AC	2 CONDUCTOR CORD, RUBBER-JACKETED AWG 18 (41 × 34) .245 O.D.
5	1	J1	JACK	
4			CAPACITOR ELECTROLYTIC	
3	1		FUSE	
2	✕		SCHEMATIC DIAGRAM	EMD-1 Sheet 1 of 1
1	✕		WIRING DIAGRAM	WD-2 Sheet 1 of 1 (Page 63)

LIST OF MATERIAL

ITEM	NO. REQ'D	DESCRIPTION	MANUFACTURER & PART NO. OR MIL TYPE DESIGNATION
15	6	WASHER, LOCK, SPLIT, NO. 8	
14	4	WASHER,	MS35337-78
13	6	WASHER,	MS15795-307
12	4	WASHER, FLAT, NO. 4, CRES	
11			MS35649-84
10	4		MS35649-44
9	4	SCREW, PAN HD. NO. 8-32 × ⅜ LONG	
8	2		MS35221-44
7	4		MS35221-18
6	1	SOLDER LUG (NO. 8)	
5	1		
4			
3	1		LITTELFUSE, INC. NO. 342001
2	1	TERMINAL BOARD ASSEMBLY (TB1)	EMD-2 SHEET 2
1		CHASSIS	EMD-3 SHEET 1

TITLE ELECTROMECHANICAL DESIGN—ELECTRICAL & MECHANICAL PARTS LIST

SCALE

GRADE

COURSE

DATE

NAME

INTRODUCTION: THE WIRING DIAGRAM

A **wiring diagram** (or connection diagram) shows pictorially, or in list form, the **wire connections** of an electronic assembly or its components. These connections may be external, internal, or both; however, the external-connection drawing is usually referred to as an **interconnection diagram.** An example shown in Fig. 1.

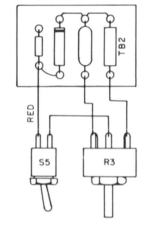

RED

S5

R3

TB2

FIG. 1

Exercise. Shown in Fig. 2 is the top view of the **power supply** in lesson 5, SCH-1 sheet 1 (page 25). Fig. 3 is the bottom view of the power supply.

Redraw the bottom view of the power supply and show the wiring connections between all the components. For reference see schematic diagram of the power supply in SCH-1.

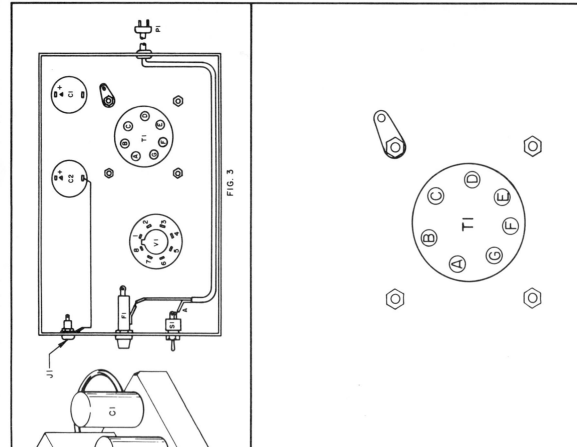

T1

C1

C2

V1

P1

FIG. 2

P1

C1

C2

T1

D

C

B

A

E

F

G

F1

S1

A

J1

FIG. 3

T1

A B C D E F G

TITLE	WIRING DIAGRAM–INTRODUCTION		DWG. NO. **WD–1**
NAME	DATE		PAGE 62
	COURSE	GRADE	SCALE NONE	SHEET 1 OF 1

Exercise. Complete the started wiring diagram and wiring list of the power supply that you designed in the lesson on electromechanical design. Reference designations of the components should conform to those in the schematic diagram of the power supply (EMD-1, sheet 1, page 55). Notice that **blocks** are being used to represent component outlines. The **wiring list**, which is self-explanatory here, is commonly used in very complicated wiring diagrams.

LENGTH INCHES	AWG	COLOR	ITEM NO.	FROM LOCATION	TO LOCATION
6	22	BLACK	1	P1	F1
			2		S1
		G/B	3	T1	F1
			4	T1	TB1-1
9		G	5	T1	
		R	6	T1	
		G	7	TB1-6	
		RED	8	TB1-11	+C1
			9	TB1-12	
		BLACK	10	C1	
		BLACK	11	C2	
			12	J1	GND.LUG
	22		13	+C2	

ALL WIRES #22 AWG.

TITLE: WIRING DIAGRAM-POWER SUPPLY

NAME   DATE   COURSE   GRADE   SCALE NONE   DWG. NO. WD-2

SHEET 1 OF 1

INTRODUCTION: THE INTERCONNECTION DIAGRAM

An **interconnection diagram** is a drawing which shows the external wiring connections between items of equipment or unit assemblies. The internal connections of the unit assemblies are generally omitted.

In Fig. 1 is shown a typical audio system consisting of five units:

UNIT NO. 1: Record player
UNIT NO. 2: Tape recorder
UNIT NO. 3: Preamp
UNIT NO. 4: Power amplifier
UNIT NO. 5: Speaker

All these units are interconnected with cables numbered from W1 through W4. This audio system can be shown drawn as an **interconnection diagram** by drawing each unit as a **block** and each cable as a line, thus eliminating pictorial drawings.

A **unit** is identified by a unit number such as 1, 2, 3, etc.

A **cable** is identified with a W such as W1, W2, W3, etc.

The unit will always show the connector (jack) marked J1, J2, etc. Thus 2J1 means UNIT NO. 2, JACK NO. J1. The cables will be identified as shown:

Thus W1P1 means CABLE NO. W1, CONNECTOR NO. P1.

FIG. NO. 1

Exercise. Complete the **interconnection diagram** started below of the audio system (Fig. 1) by showing a single-line drawing identifying all **blocks, units, connectors,** and **cables.**

RECORD PLAYER J1 P1 W1
UNIT NO. 1

TITLE	INTERCONNECTION DIAGRAM–AUDIO SYSTEM			DWG. NO. ID–1	
NAME	DATE	COURSE	GRADE	SCALE NONE	PAGE 64
					SHEET 1 OF 1

Exercise. Draw an **interconnection diagram** of a digital computer system and a list of routing of all cables without the help of a sketch. In more complicated interconnection diagrams, a list of cable routing is extremely helpful. The digital computer consists of the following nine units:

UNIT NO. 1: Arithmetic unit (6 jacks, J1 through J6)
UNIT NO. 2: Typewriter (1 jack, J1)
UNIT NO. 3: Tape recorder (1 jack, J1)
UNIT NO. 4: Card reader (1 jack, J1)
UNIT NO. 5: Plotter (1 jack, J1)
UNIT NO. 6: Power supply (1 jack, J1)
UNIT NO. 7: Core memory (1 jack, J1)
UNIT NO. 8: Disk storage (1 jack, J1)
UNIT NO. 9: Tape storage (1 jack, J1)

All the **units** from UNIT NO. 2 through 9 have one single **jack** (J1). UNIT NO. 1 has 6 JACKS numbered from J1 through J6 as shown. Show cable routing as follows;

From UNIT NO. 1 to UNIT NO. 2 J1 (CABLE NO. W1)
                    UNIT NO. 3 J1 (CABLE NO. W2)
                    UNIT NO. 4 J1 (CABLE NO. W3)
                    UNIT NO. 5 J1 (CABLE NO. W4)
                    UNIT NO. 6 J1 (CABLE NO. W5)
                    UNIT NO. 7 J1
                    UNIT NO. 8 J1 } (ONE SINGLE
From UNIT NO. 1 to UNIT NO. 9 J1   CABLE W6)

Complete the CABLE ROUTING LIST.

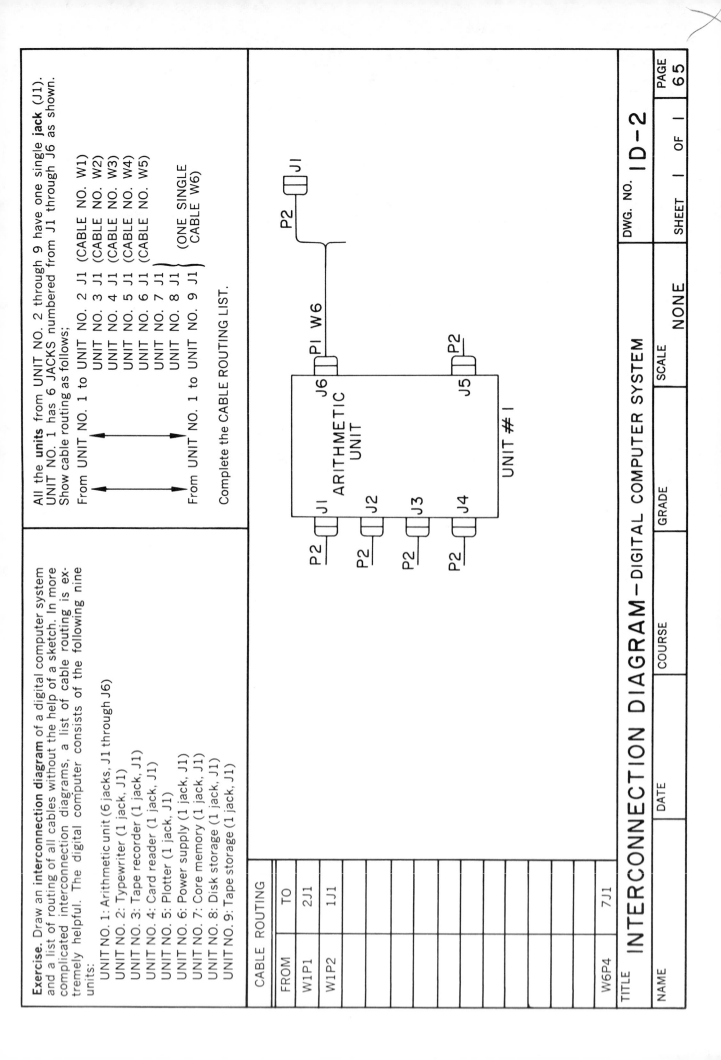

CABLE ROUTING

FROM	TO
W1P1	2J1
W1P2	1J1
W6P4	7J1

TITLE INTERCONNECTION DIAGRAM – DIGITAL COMPUTER SYSTEM

NAME	DATE	COURSE	GRADE	SCALE NONE	DWG. NO. ID-2	PAGE 65
					SHEET 1 OF 1	

PICTORIALS

In most large companies, a technical illustrator does the "formal" **pictorials.** Smaller companies have their own draftsmen do them. But perhaps the most widespread function of the pictorial is its use in sketches between draftsmen, designers, and engineers to develop ideas. There are many types of pictorials. Probably the most common and easiest to draw are the **cabinet and isometric drawings.** These are illustrated at right. Neither the cabinet nor the isometric drawing is an accurate picture. Therefore a certain amount of "fudging" (guesswork and inaccurate touch-up) is acceptable.

Exercise 1. Draw a second bracket on cabinet and isometric views (top and middle).

Exercise 2. In the isometric boxes at bottom right draw three 8-point circles in each surface center of box A and three 4-point center circles in each surface center of box B. Make all holes ½" diameter.

REGULAR MULTIVIEW DWG.
(ORTHOGRAPHIC PROJECTION)

Cabinet drawings are basically made with **full scale length and height** but **half-scale depth,** with 45° **receding lines.** Fudging as shown reduces distortion.

FULL SCALE
(NO FUDGING)

½ SCALE
(NO FUDGING)

½ SCALE FOR RECEDING SMALL DIMS.

¾ SCALE (FUDGED)

FULL SCALE HEIGHT

45°

½ SCALE THICKNESS OR DEPTH (SHORTEST DIM.)

SCALE (FUDGED)

¾ SCALE (FUDGED)

FULL SCALE LENGTH

ISOMETRIC HOLES, 4-HOLE METHOD

(1) LOCATE CENTER

(2) BOX DIA.

(3) MAKE CONSTR. LINES

(4) DRAW ARCS FROM ARC CTRS

4-POINT CENTER CIRCLE METHOD

ISOMETRIC CIRCLE, 8 POINTS

HINT: LOCATE EACH OF THE 8 POINTS ISOMETRICALLY

FULL SCALE

FULL SCALE

FULL SCALE

30°

30°

90°

8-POINT CIRCLES

A

B

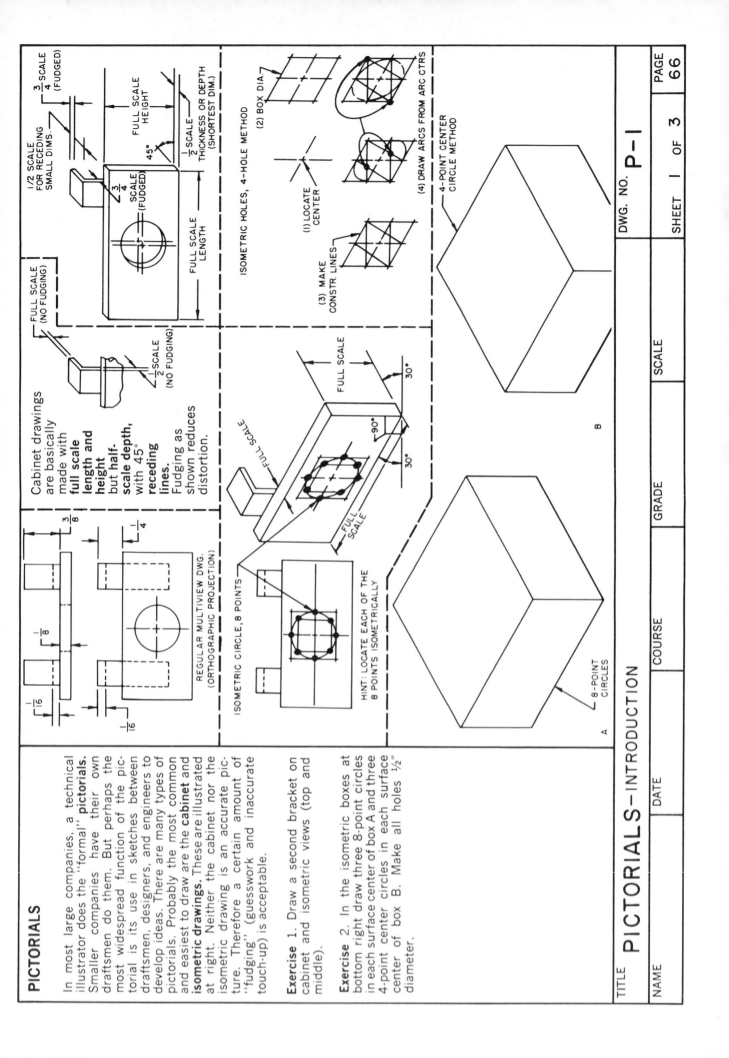

TITLE PICTORIALS—INTRODUCTION

DWG. NO. P-1

PAGE 66

SHEET 1 OF 3

NAME

DATE

COURSE

GRADE

SCALE

From sheet 1 it was apparent that **cabinet drawings** lend themselves to relatively thin objects, while the boxlike subject shows up better in **isometric drawings.**

Exercise. Complete the cabinet drawing started below of the **terminal board assembly,** 2 X size. For reference see Electromechanical Design (lesson 9) sheets 1 and 2, pages 57 and 58. The terminals are USECO No. 2000B (Appendix C, page 96).

TITLE	PICTORIALS–CABINET DRAWING–TERMINAL BOARD			DWG. NO. P-1	
NAME	DATE	COURSE	GRADE	SCALE 2/1	PAGE 67
					SHEET 2 OF 3

In **isometric drawings,** a cylinder can be drawn like a hole. Any curved line can be regarded as a portion of a circle and may be located by a series of points. In using a template (ellipse) for ISO-METRIC CIRCLES or curves, be sure to properly align the **major axis** of the ellipse so that it is 90° or at right angles to the **hole axis** (Fig. 1).

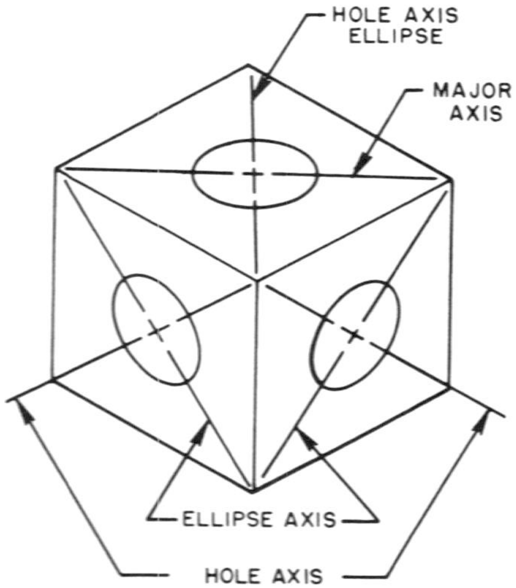

HOLE AXIS
ELLIPSE

MAJOR
AXIS

ELLIPSE AXIS

HOLE AXIS

Fig. 1

Exercise. Complete the **isometric** below. Draw the power supply of lesson 9, EMD-4 sheet 1 (page 60).

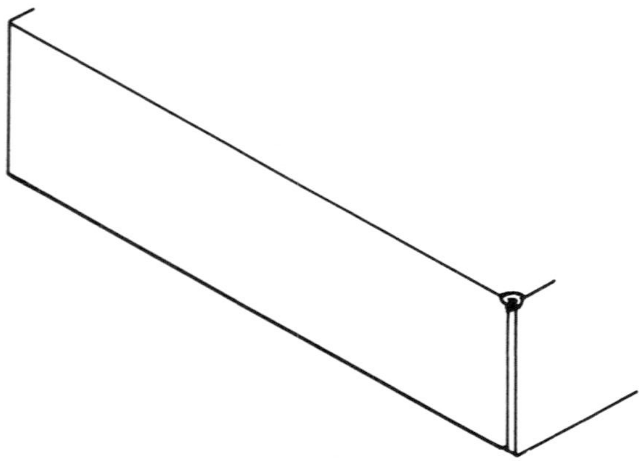

SCALE: HALF SIZE

TITLE	PICTORIALS—ISOMETRIC DRAWING			DWG. NO. P-1	
NAME	DATE	COURSE	GRADE	SCALE 1/2	PAGE 68
					SHEET 3 OF 3

Note: Three-Quarter size template MIL-STD-806 will be required for the next exercises (page 70).

LOGIC TEMPLATE. The standard logic symbol template No. MIL-STD-806 shown at right is used to draw logic symbols in much the same manner as the USAS Y32.2 template is used to draw Electronic Schematic Symbols.

The MIL-STD-806 is often referred to as ASA Y32.14 since they are one and the same.

To familiarize yourself with the symbols, study Appendix F (pages 103 and 104) and do the exercises below.

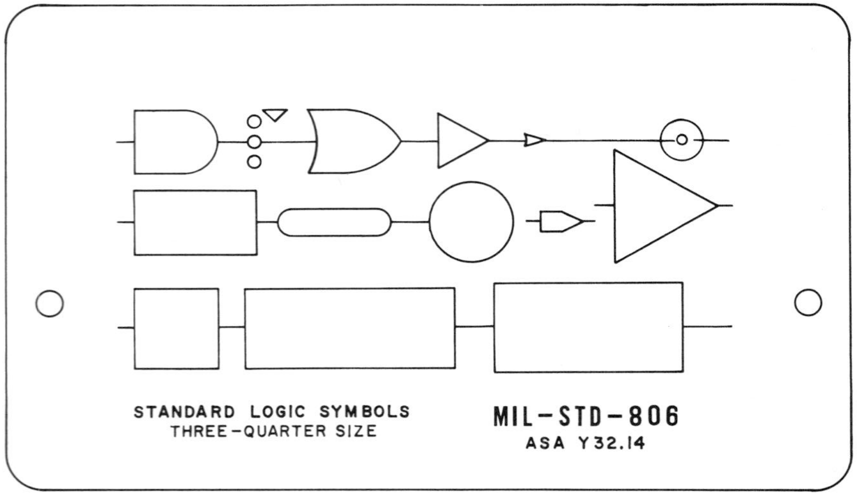

1 What is a flip-flop? (See explanation, page 103.)

THE FLIP-FLOP IS _____

_____

2 Complete the statement: "THE OR OUTPUT IS HIGH (H) IF AND ONLY IF (page 103)

_____

3 If the INCLUSIVE OR inputs were: A = **HIGH**, B = **LOW**, C = **LOW**; THE OUTPUT F WOULD BE

_____

4 If the INCLUSIVE OR function above (Problem 3) had a **small circle** attached to the output, the output F would be _____

5 What **MIL** specifications describe the template shown above? _____ What is an equivalent code specification? _____

6 The **time-delay** symbol of Appendix F describes duration of delay in **milliseconds**, while the **single-shot function** describes **output duration** in _____

7 The aspect ratio (page 104) of the **general logic** symbol is _____

LOGIC TEMPLATE MIL-STD-806, THREE-QUARTER SIZE

STANDARD LOGIC SYMBOLS
THREE-QUARTER SIZE

MIL - STD - 806
ASA Y 32.14

TITLE	LOGIC SYMBOL FAMILIARIZATION			DWG. NO. LOG-I	
NAME	DATE	COURSE	GRADE	SCALE	PAGE
				3/4 SIZE	69
				SHEET I OF 2	

With MIL-STD-806 template, fill in the spaces below. Three symbols and titles per line.

SHIFT REGISTER

AND

INCLUSIVE OR

SCHMITT TRIGGER

EXCLUSIVE OR

FLIP-FLOP

SINGLE SHOT

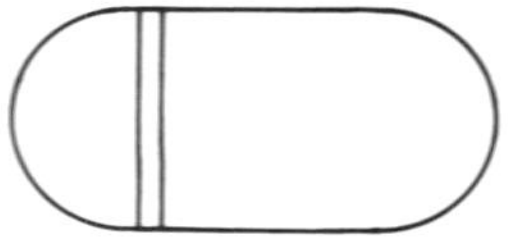

TIME DELAY

AMPLIFIER

TITLE	LOGIC SYMBOLS – TEMPLATE FAMILIARIZATION		DWG. NO. LOG–I	
NAME	DATE	COURSE	GRADE	SCALE 3/4 SIZE
				PAGE 70
				SHEET 2 OF 2

With MIL-STD-806 template, complete below the started logic diagram in Appendix F, page 105.

TITLE LOGIC DIAGRAM			DWG. NO. LOG-2	
NAME				PAGE 71
	DATE	COURSE	GRADE	SCALE
				3/4 SIZE	SHEET 1 OF 1

INTRODUCTION TO INTEGRATED CIRCUITS (I.C.)

Integrated circuits are combinations of active elements (such as diodes and transistors) and passive elements (such as capacitors and resistors) functioning as a single item. They generally are made as very small wafers, referred to as "chips" or "dies" 1/16 inch square or even smaller. Within this tiny area, an integrated circuit may contain the equivalent of 50 components or more. Two or more of these circuits are commonly encapsulated in flat rectangular packages known as "flat-packs."

At present there are two principal types of I.C. packages (or chips); (a) the 14- or 16-pin **dual-in-line** as shown on page 106 of Appendix G, Fig. 2 (also called the "bug") and (b) the 10- or 14-pin "flat pack" shown on page 107 of Appendix H, Fig. 1 and 2.

The dual-in-line has become more commonly used than the flat pack and requires plated-thru holes (see page 106, Fig. 1). The flat pack is mounted with its leads above the board (see page 107, Fig. 3 through Fig. 3a). It is recommended that all "feed thru" connections on all I.C. boards (both flat pack and dual-in-line) use plated-thru holes rather than buss wire or eyelets.

Both the dual-in-line and flat pack chip utilize a corner **black dot** for orientation purposes; these are illustrated on pages 106 and 107. In particular the dual-in-line dot is located nearest pin 1 and the other pins are ascertained by counting **counter-clockwise** from pin 1.

Appendix G, Fig. 3 shows the internal logic of the various chips, the components of which are known as elements of the I.C.

Complete the exercise at right and continue with page 73. In the lessons that follow, further explanations will be made in greater detail regarding the use of the flat pack and dual-in-line with their requirements for drafting and design applications.

Exercises (Fill in the blanks)

1 Integrated circuits are generally made into small wafers referred to as _____ or _____ .

2 The 10-pin I.C. package is known as a _____ while the 16-pin I.C. package is known as a _____ .

3 What kind of "feed-thru" connections should a designer select for I.C. boards? _____ .

4 The black dot on the dual-in-line package lies nearest pin _____ .

5 In locating pin 10, the user should start with the black dot and count in a _____ direction.

6 The flat pack comes in either _____ pins or _____ pins while the dual-in-line comes in either _____ or _____ pins.

7 The "bug" is another name for the _____ package.

8 The package having leads mounted above the board is called the _____ while the package requiring feed-thru holes is called the _____ .

TITLE	INTRODUCTION TO INTEGRATED CIRCUITS (I.C.)			DWG. NO. ICI-1	
NAME	DATE	COURSE	GRADE	SCALE	PAGE
					72
				SHEET 1 OF 1	

14-Pin Dual-in-Line

Fig. 2 of Appendix G (page 106) illustrates the 14-pin dual-in-line I.C. Notice that the leads are bent downward and in line for easy assembly onto P.C. boards. The dual-in-line leads are inserted into plated-thru holes.

Exercise: Draw three views (front, top, and side) of the 14-pin dual-in-line 2 X size on the right (see page 106). Show all the dimensions, the black dot, and number all the leads as started in Fig. 2. After completing the drawing, fill in the blanks below.

1 The length of the dual-in-line is _____ inches.

2 The width of the dual-in-line is _____ inches.

3 The diameter of the dual-in-line leads is _____ inches.

4 The black dot is located near pin _____ .

TITLE	14–PIN DUAL–IN–LINE PACKAGE INTEGRATED CIRCUIT (I.C.) OUTLINE		DWG. NO. IC-2	
NAME	DATE	COURSE	GRADE	SCALE 2 X SIZE

SHEET 1 OF 1

Logic diagrams, such as the one drawn on page 71, are made up of several integrated circuit symbols, designated by the letter "U."

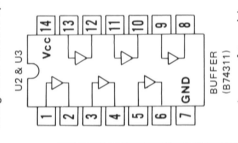

U2 & U3

BUFFER
(B74311)

Each "U" symbol represents some "functional elements" **inside** a chip or integrated circuit (I.C.). The chip shown at left is made up of six elements.

The element shown ⊳ is called a "buffer."

From page 71 we can count eleven buffers, six in U3 and five in U2. From the chip shown on left we can count six buffers each for U2 and U3 totaling twelve buffers. This implies that Diagram (page 71) is allowing one buffer to remain unused – spare elements are quite common. An example of unused elements can be seen by noting that on page 106, the I.C. called "flip-flop" (F74921) is made up of two identical elements: if only one element is used, as on page 71, it will be denoted as "1/2 of U5."

Each I.C. chip has a ground pin (GND = pin 7) and a voltage pin (Vcc = pin 14) as shown above; the remaining pins are represented on diagrams for example as:

or

Logic diagrams are drawn from left (input) to right (output). Page 71 has encircled numbers ② on the left and right sides representing **board terminals** or **P.C. tabs** – not chip pins.

Chips are purchased by part numbers such as B74311 or F74921.

Exercise: Use the diagram on page 71 and the I.C. chips on page 106 and complete the following:

1. What letter is used on logic diagrams to represent an I.C.? _____

2. From page 71, how many element symbols are used in U1? _____

3. How many functional elements are in the quad gate, Q74513 I.C. chip? _____

4. What fraction of the chip U5 is being used? _____

5. On all 14-pin chips, pin 7 is used for _____.

6. On all 14-pin chips, pin 14 is used for _____.

7. On page 71, pin 7 of all I.C. are connected to terminal number _____.

8. Chip 1/2 of U5, pin 6 is connected to terminal number _____.

9. U3 pin 10 is connected to U4 pin _____.

10. How many functional elements are in the flip-flop chip, F74921? _____

11. Complete the drawing on the right of the buffer chip (B74311). Show all elements and pin numbers.

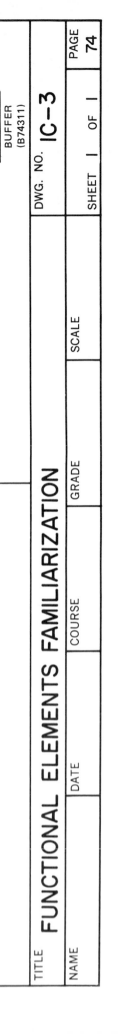

BUFFER
(B74311)

TITLE	FUNCTIONAL ELEMENTS FAMILIARIZATION		DWG. NO. IC-3
NAME	DATE	COURSE	GRADE	SCALE

PAGE 74

SHEET 1 OF 1

Integrated Circuit (I.C.) Design Layout Requirements

To layout an I.C. board, the designer needs the following items: (1) a logic diagram; (2) a board outline; and (3) .100 quadrille grid mylar or graph paper.

I.C. boards should be drawn 2 × size accurately on .100 quadrille grid mylar or graph paper as shown in Fig. A. The layout is viewed from top component side (see page 106, Fig. 1 for reference) showing all the I.C. dual-in-line pads (1/8 O.D. — 2 × size) without components. Circuitry conductors are shown in the usual way:

      Component (top) side = solid lines (——————)

      Circuit (bottom) side = dashed lines (— — — — —)

Fig. A illustrates how long and short conductor lines are drawn. Short conductor lines are used to connect long lines to pads, plated-thru holes or P.C. tabs. It is recommended that long conductor lines be drawn as follows:

      Top solid long lines — horizontally (Except for GND and Vcc lines — see page 76A)

      Bottom dashed long lines — vertically

Plated-thru holes are used when it is necessary to connect a top conductor line to a bottom conductor line. Plated-thru holes, other than I.C. pads, are indicated thus:

      ◉ (1/8 O.D. = 2 × size)

In making the layout of the I.C. board, **do not** draw the outline of the I.C. chip — it is not needed.

The hole pattern of I.C. pads and plated-thru holes are located on **grid intersections** as shown in Fig. A and Fig. B.

The conductor line widths need not be drawn to scale, however the line widths for ground (GND) and voltage (Vcc) should be heavy — .100 inch (2 × size).

The short conductor line connections to I.C. pin 7 and 14 pads may be drawn narrower than .100 inch; see Fig. A.

FIGURE A

DUAL-IN-LINE (14 pins)
I.C. PADS (1/8 DIA)
.100 GRID

BOARD OUTLINE

VERTICAL **LONG** CONDUCTOR LINES–DASHED ON BOTTOM

PLATED-THRU HOLE 1/8 O.D.

.100 —| |— BETWEEN LINES

Vcc
.100 THK (2 × SIZE)

SHORT CONDUCTOR LINES (BOTTOM)

SHORT CONDUCTOR LINE (TOP)

GND
.100 THK (2 × SIZE)

HORIZONTAL LONG CONDUCTOR LINE— SOLID ON TOP

1/8 DIA (2 × SIZE)
.200 (2 × SIZE) BETWEEN PADS Ҫ TO Ҫ

FIGURE B

Exercise:

1 The three major items needed to lay out an I.C. board are:

a. _____
b. _____
c. _____

2 The quadrille grid lines on mylar or graph paper are spaced _____ apart.

3 What symbol is used to denote a plated-thru hole connecting top and bottom conductors of the board on a layout? _____

4 Line width for GND and Vcc (2 × size) is _____.

5 The hole pattern of I.C. pads and plated-thru holes are located on _____.

6 In making the layout of the I.C. board, (do/do not) _____ draw the outline of the I.C. chip.

7 The I.C. pad diameter is _____ (2 × size).

8 Circuitry conductors shown on component (top) side = _____ lines (_____); on the circuit (bottom) side = _____ lines (_____).

TITLE	INTEGRATED CIRCUIT (I.C.) DESIGN LAYOUT REQUIREMENTS			DWG. NO. IC-4	PAGE
NAME	DATE	COURSE	GRADE	SCALE 2 × SIZE	SHEET 1 OF 1

DUAL-IN-LINE LAYOUT TECHNIQUE

To prepare a layout, using dual-in-line chips, proceed as follows, using the illustration at left for reference.

Start by placing all I.C. chips in a vertical position (see phantom view – but do **not** draw chip outline) parallel with the tabs and with the pin 1 of each chip located at the left corner. The spacing between chips can vary depending on the number of chips being used. This method of layout gives the board a uniform, neat appearance and lends itself to ease of assembly in production.

Pin 1 pad is the only pad of each chip that needs to be labeled; then locate the 14 pads of each chip. Label the position of each I.C. chip with a "U" number such as "U1," "U2," "U3," etc. The rough layout can be filled in now by showing all the conductors and plated-thru holes using the logic diagram as a reference.

To complete the "finishing touches", do the ground (GND) and voltage (Vcc) conductor lines first. These lines are placed on top of the board with the GND lines proceeding from the left GND tab and the Vcc lines proceeding from the right Vcc tab as shown. Leave one grid edge spacing around the board outline and between conductor lines (.100).

Both solid and dashed lines should follow the grid pattern and all bends should be made with 90 degree corners. Plated-thru holes should **not** be located under an I.C. – they should be visible for inspection purposes from both sides of the board.

Exercise: Complete the following:

1. I.C. chips should be placed in a _____ position, parallel with the tabs.

2. The I.C. Chip is oriented so that the pin 1 is located in the _____ corner.

3. The GND conductor originates from a tab located on the _____ side of the board.

4. The Vcc conductor originates from a tab located on the _____ side of the board.

5. Conductor lines should be bent with _____ degree corners.

REF DESIGNATION

SPACING BETWEEN CHIPS CAN VARY

PIN 1
(UPPER LEFT)

Vcc
RIGHT SIDE

I.C. CHIP
DO NOT DRAW
OUTLINE ON
LAYOUT

U1   U3

U2   U4

GND
LEFT SIDE

.030 THK
(2 × SIZE)

TABS

GND

VCC

1 2 3 4 5 6

.100 OR
ONE GRID
SPACING

TITLE	DUAL-IN-LINE LAYOUT TECHNIQUE			DWG. NO.	IC-5
NAME	DATE	COURSE	GRADE	SCALE 2 X SIZE	PAGE 76A
				SHEET 1 OF 2	

Exercise:

Study the logic diagram below, and complete the started layout at right by filling in the missing conductor lines.

First. Connect all GND conductors on the pads for pins 7 of all the I.C. chips to tab 1 in the same manner as was shown in the previous exercise (page 76A).

Second. Follow the same procedure of page 76A to connect all the Vcc conductors on the pads for pins 14 of all the I.C. chips to tab 6.

The last step is to connect **four missing** conductor lines as per the logic diagram; it probably is simpler to proceed from left to right. (Hint: one of the four missing lines will require a plated-thru hole).

4 DUAL-IN-LINE LAYOUT
(2 × SIZE)

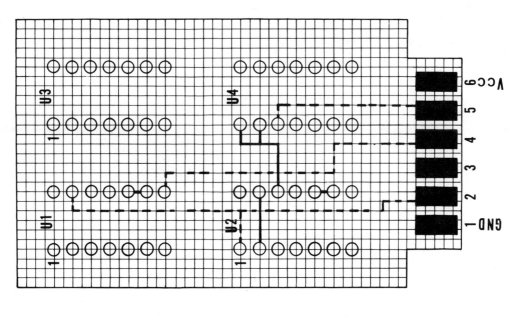

LOGIC DIAGRAM

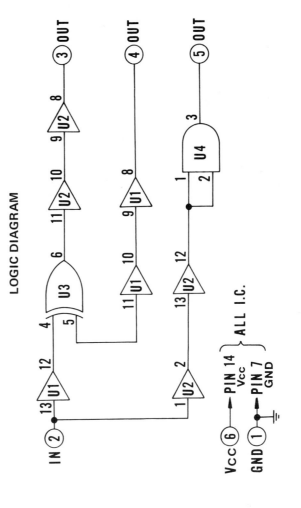

TITLE	DUEL–IN–LINE LAYOUT TECHNIQUE				DWG. NO. IC-5	PAGE 76B
NAME		DATE	GRADE	COURSE		
					SCALE 2 X SIZE	SHEET 2 OF 2

Exercise:

Complete the I.C. board design layout started at right. To complete the layout, use the logic diagram drawn on page 71 and the I.C. chips of page 106 (Fig. 3).

Notice that there are five 14-pin dual-in-line chips mounted on the board (U1 through U5).

The board is drawn $2 \times$ size on a .100 Grid.

The tabs are numbered from 1 through 14, corresponding to the logic diagram inputs and outputs. The same numbered tabs are both on top and bottom of the board and conductors can extend from either or both sides of the tab.

Recall that on page 76B you started your layout by drawing in the GND and Vcc conductor lines first (on top); this procedure should be followed again in the layout on your right.

Use the logic diagram (page 71) as a reference to complete the layout, showing all solid and dashed lines plus all plated-thru holes as required.

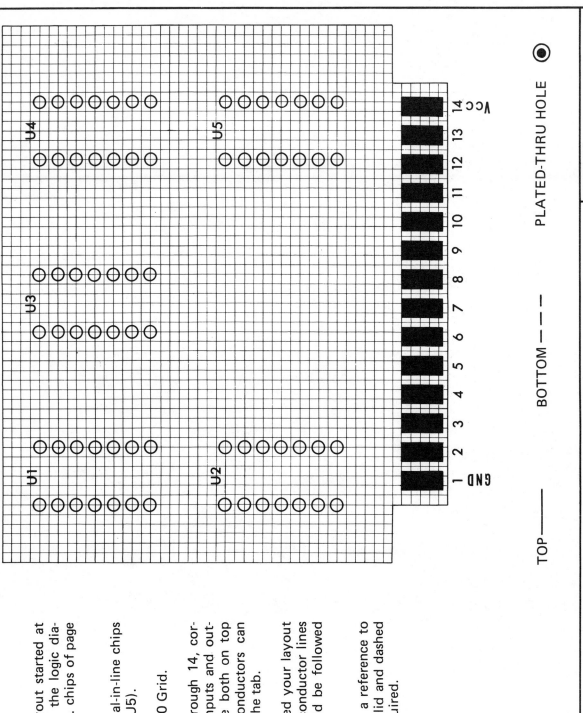

PLATED-THRU HOLE ◉

TOP ———

BOTTOM — — —

TITLE		I.C. BOARD DESIGN LAYOUT — 5 DUAL-IN-LINE		DWG. NO.	IC–6	
NAME	DATE	COURSE	GRADE	SCALE 2 X SIZE		PAGE 77
				SHEET 1 OF 1		

The 14-Pin Flat-Package I.C. Fig. 2 of Appendix H (page 107) illustrates the 14-pin flat-package. The flat-package is mounted with its leads above the board (see page 107, Fig. 3).

Exercise: Draw the front and top views of the 14-pin flat-package (page 107, Fig. 2) 10 x size below. Label the pins 1, 2, 3, etc., as shown with the black dot between numbers 1 & 2.

Answer the following. The 14-pin flat-package has a:

1 leads length of _____ inches.

2 leads width of _____ inches.

TITLE	14–PIN FLAT–PACKAGE INTEGRATED CIRCUIT (I.C.) OUTLINE		DWG. NO. IC–7	PAGE 78
			SCALE	SHEET 1 OF 1
			10 X SIZE	
NAME	DATE	COURSE	GRADE	

I.C. HOOK-UP, DISTRIBUTION OF LEADS — FLAT PACKAGE

Integrated Circuit (I.C.) packages are often layed out in rows with certain common leads hooked "in parallel." For example, in most 14-pin flat-package the **power** lead is pin **4** (Vcc); this pin (or lead) is usually connected in parallel to pin **4** leads of the adjacent flat-packages. Similarly, all the **ground leads** or pin **10** (GND) are connected in parallel.

In the layout at right, the parallel connection of all the **4** pins has been started. Similarly, all the **10** pins are connected in parallel as are all the **input leads** or pin **3**. Each set of pins is **separately** hooked in parallel. The **4** pins and **10** pins have been laid out with connections on the **top side** of the board; the **3** pin connections are made on the **bottom side** via **plated-thru holes**.

The remaining unused leads are normally soldered to the nearest **blank** (unconnected) **lap joint** or **land** (see Appendix H, Fig. 3.).

Exercise 1:

a. Which number is used to represent the **power lead** (Vcc)? _____

b. Which number is used to represent the **ground lead** (GND)? _____

c. Which number is used to represent the **input lead?** _____

Exercise 2: Connect the flat-packages A1, A2, & A3 below so that the pins **3**, **10**, & **4** are each separately hooked in parallel. Draw in all leads & cross-hatching.

Key: Top side conductors: ▨▨▨

Bottom (hidden) side conductors: – – – – –

.025 FULL SCALE ←→

.020 DIA. HOLE, PLATED THRU
FOR 4 X SIZE LAYOUT USE
.250 O.D. X .080 I.D. PAD

TITLE **INTEGRATED CIRCUIT HOOK-UP** (FLAT PACKAGE)

NAME	DATE	COURSE	GRADE	SCALE 4X SIZE	DWG. NO. IC-8	PAGE 79
					SHEET 1 OF 1	

COMPONENT SYMBOLS

COMPONENT	SYMBOL	REF. DESIG.	APPENDIX B PAGE NO.
AMPLIFIER Two inputs		AR	
ANTENNA General Dipole		E	
BATTERY One cell Multicell		BT	
CAPACITOR General		C	85
Polarized (electrolytic)			86, 94
Variable or Adjustable			
CONDUCTOR Wiring (general) Cable, 1 or more conductors			

COMPONENT	SYMBOL	REF. DESIG.	APPENDIX B PAGE NO.
Crossing not connected			
Junction (avoid if possible. Use next symbol)			
Junction			
CONNECTOR Male pin Female Socket		P J	
Separable connectors (engaged)			
Engaged 4-conductor connectors. Letters or numerals need not be alphabetical or numerical in order.			
Coaxial, connector, build-up example. R-F connector (plug)		P P	

APPENDIX A

COMPONENT SYMBOLS

COMPONENT	SYMBOL	REF. DESIG.	APPENDIX B PAGE NO.
CONNECTOR Mating coaxial			
Two-conductor (Jack)		J	
Two-conductor (Plug)		P	
Power, female contacts (2 conductors)			
Power, male contacts			
CRYSTAL Piezoelectric		Y	93
FUSE		F	93
GROUND Wire is bonded to chassis.			
Chassis or frame connection.			

COMPONENT	SYMBOL	REF. DESIG.	APPENDIX B PAGE NO.
INDUCTOR General		L	
Magnetic core			
Tapped			
Adjustable			
LAMP General		DS	
AC type, neon			
Incandescent filament			
MACHINE, ROTATING General		G	
Generator			
Motor		B	
MICROPHONE		MK	

APPENDIX A

APPENDIX A

COMPONENT SYMBOLS

COMPONENT	SYMBOL	REF. DESIG.	APPENDIX B PAGE NO.
RELAY		K	
Contactor (old style)			
Contactor			
Double-pole double-throw (DPDT) (old style)			
Double-pole double-throw (DPDT)			

COMPONENT	SYMBOL	REF. DESIG.	APPENDIX B PAGE NO.
METER (INSTRUMENTS) General		M	
Example: milliammeter			
RECTIFIER Semiconductor rectifier diode		CR	87
Breakdown diode, unidirectional			87
Breakdown diode, bidirectional			87
Tunnel diode			

COMPONENT	SYMBOL	REF. DESIG.	APPENDIX B PAGE NO.	COMPONENT	SYMBOL	REF. DESIG.	APPENDIX B PAGE NO.
RESISTOR General	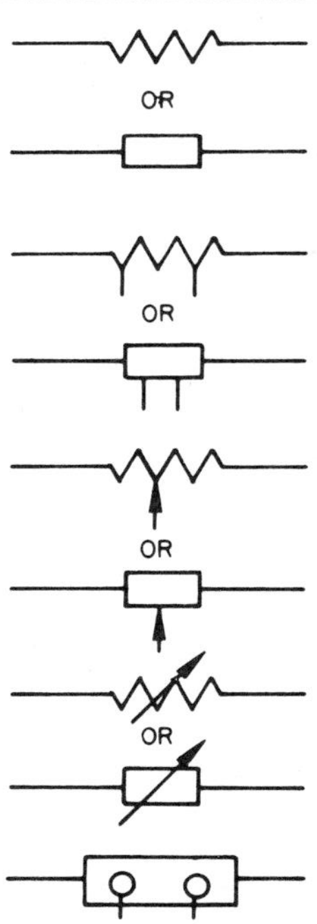	R	88, 89	SWITCH Single-Throw (ST)		S	90, 95
Tapped			90	Double-throw (DT)			
Adjustable				Double-pole, Double-throw (DPDT)			
Variable				Push-button			
Shunt				Selector			
				Wafer, 3 poles, 3 circuits with 2 nonshorting and 1 shorting moving contacts.			
SHIELDED CONDUCTOR Single conductor				TERMINAL BOARD Group of 4 terminals		TB	
Five-conductor cable, shield grounded				TRANSFORMER General		T	93
				Magnetic core			
SPEAKER General		LS		Tapped			

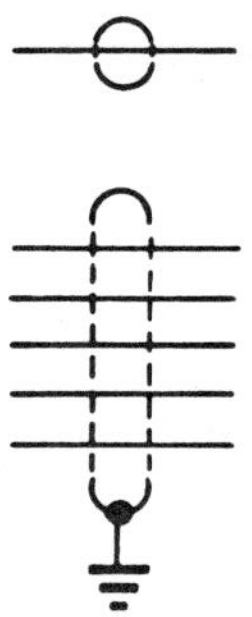

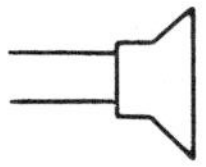

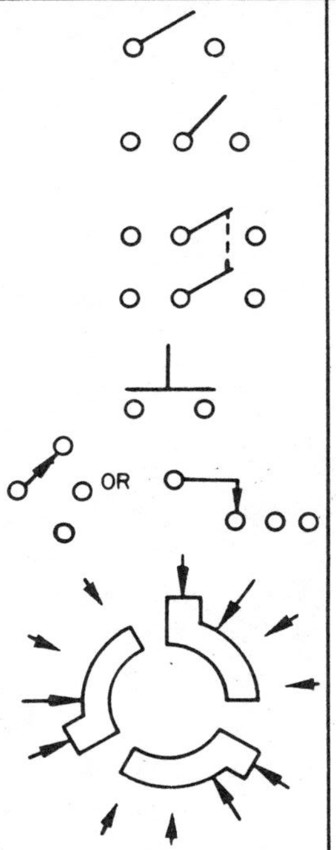

COMPONENT SYMBOLS

APPENDIX A

APPENDIX A

COMPONENT	SYMBOL	REF. DESIG.	APPENDIX B PAGE NO.
TUBE Triode		V	92
Twin triode			
TUBE ENVELOPE General	OR		
Split			
Gas-filled			
C.R.T.			
WAVEGUIDE Circular		W	
Rectangular			

COMPONENT	SYMBOL	REF. DESIG.	APPENDIX B PAGE NO.
TRANSISTOR PNP type		Q	91 ←——————————→ 91
PNP type with 1 electrode connected to envelope			
NPN type			
Unijunction N-type base			
Unijunction P-type base			
Field-effect N-type base			

COMPONENT SYMBOLS

Mica Capacitors

COMPONENT: CAPACITOR, MOLDED SILVERED MICA, 500 VDC, ±5% (MIL TYPE CM-15).

MANUFACTURER: ARCO ELECTRONICS, INC.

CAP. IN MMF.	MIL TYPE DESIGNATION	CAP. IN MMF.	MIL TYPE DESIGNATION
1	CM-15-C-010J	91	CM-15-E-910J
2	-C-020J	100	-E-101J
3	-C-030J	110	-E-111J
5	-C-050J	120	-E-121J
10	-C-100J	130	-E-131J
12	-C-120J	150	-E-151J
15	-C-150J	160	-E-161J
18	-C-180J	180	-E-181J
20	-C-200J	200	-E-201J
22	-C-220J	220	-E-221J
24	-C-240J	240	-E-241J
27	-C-270J	250	-E-251J
30	-C-300J	270	-E-271J
33	-C-330J	300	-E-301J
36	-C-360J	330	-E-331J
39	-C-390J	360	-E-361J
43	-C-430J	390	-E-391J
47	-C-470J	430	-E-431J
50	-C-500J	470	-E-471J
51	-C-510J	500	-E-501J
56	-C-560J	510	-E-511J
62	-C-620J	560	-E-561J
68	-C-680J	620	-E-621J
75	-C-750J	680	-E-681J
82	CM-15-E-820J	750	-E-751J
		820	CM-15-E-821J

Ceramic CK05

COMPONENT: CAPACITOR, MICROMINIATURE, CERAMIC, 200 VDC, (MIL TYPE CK05). * = tolerance J = ±5%, K = ±10%, M = ±20%.

MANUFACTURER: VITRAMON, INC.

CAP. IN pf	MIL TYPE DESIGNATION	COMMERCIAL DESIGNATION
10	CK05CW100*	VK20CW100*
12	120*	120*
15	150*	150*
18	180*	180*
22	220*	220*
27	270*	270*
33	330*	330*
39	390*	390*
47	470*	470*
56	560*	560*
68	680*	680*
82	820*	820*
100	101*	101*
120	121*	121*
150	151*	151*
180	181*	181*
220	221*	221*
270	271*	271*
330	331*	331*
390	391*	391*
470	471*	471*
560	561*	561*
680	681*	681*
820	821*	821*
1000	CK05CW102*	VK20CW102*

Ceramic CK06

COMPONENT: CAPACITOR, MICROMINIATURE, CERAMIC, 200 VDC, (MIL TYPE CK06). * = tolerance J = ±5%, K = ±10%, M = ±20%.

MANUFACTURER: VITRAMON, INC.

CAP. IN pf	MIL TYPE DESIGNATION	COMMERCIAL DESIGNATION
1200	CK06CW122*	VK30CW122*
1500	152*	152*
1800	182*	182*
2200	222*	222*
2700	272*	272*
3300	332*	332*
3900	392*	392*
4700	472*	472*
5600	562*	562*
6800	682*	682*
8200	822*	822*
10000	CK06CW103*	VK30CW103*

Capacitor — General

COMPONENT: CAPACITOR—GENERAL PURPOSE, MIL TYPE, CM-15, CK05, CK06

NOTE: MMF IS THE SAME AS pf.

SYMBOL	REF. DESIG.
—)|(—	C

COMPONENT OUTLINE

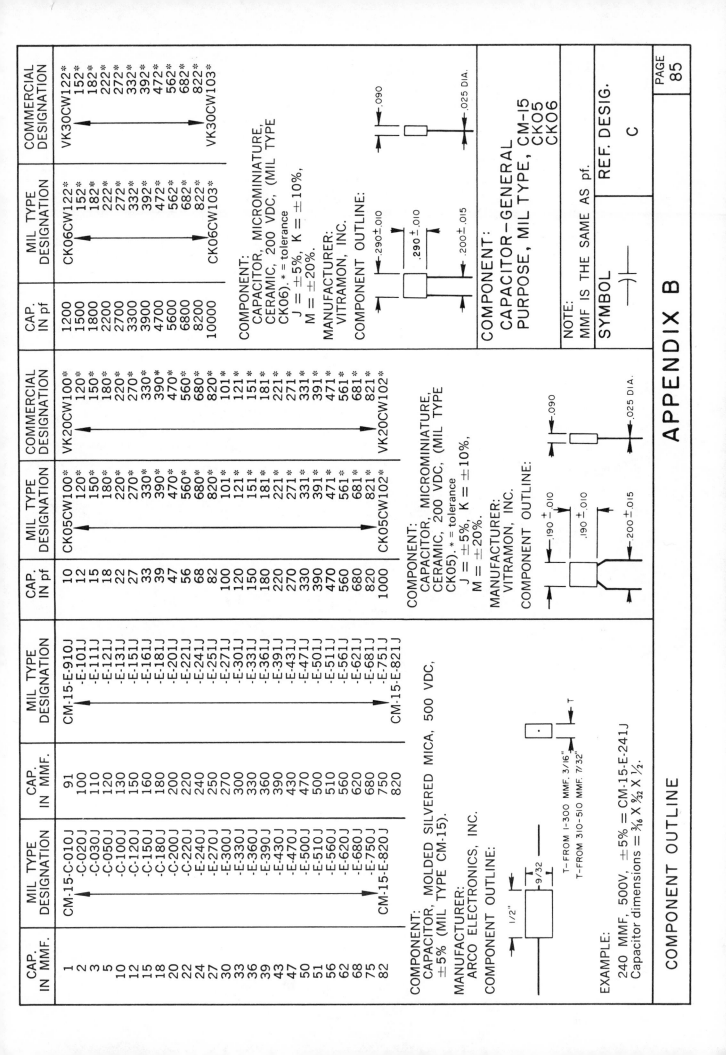

Component outline (CK06): .090, .290 ±.010, .290 ±.010, .200 ±.015, .025 DIA.

Component outline (CK05): .090, .190 ±.010, .190 ±.010, .200 ±.015, .025 DIA.

Component outline (CM-15): 1/2", 9/32, T, T-FROM 1-300 MMF. 3/16", T-FROM 310-510 MMF. 7/32"

EXAMPLE:
240 MMF, 500V, ±5% = CM-15-E-241J.
Capacitor dimensions = 3/16 X 9/32 X 1/2.

COMPONENT

CAPACITOR, FIXED, SOLID ELECTROLYTE, TANTALUM, MIL TYPE CS13 (±10%)

REF. DESIG. C

CAPACITANCE IN MFD.	DC RATED VOLTAGE	CASE SIZE	MIL TYPE DESIGNATION
5.6	6	A	CS13AB5R6K
6.8	6	A	AB6R8K
47	6	B	AB470K
56	6	B	AB560K
150	6	C	AB151K
180	6	C	AB181K
270	6	D	AB271K
330	6	D	AB331K
3.9	10	A	AC3R9K
4.7	10	A	AC4R7K
27	10	B	AC270K
33	10	B	AC330K
39	10	B	AC390K
82	10	C	AC820K
100	10	C	AC101K
120	10	C	AC121K
180	10	D	AC181K
220	10	D	AC221K
2.7	15	A	AD2R7K
3.3	15	A	AD3R3K
18	15	B	AD180K
22	15	B	AD220K
56	15	C	AD560K
68	15	C	AD680K
120	15	D	AD121K
150	15	D	AD151K
1.2	20	A	AE1R2K
1.5	20	A	CS13AE1R5K

CAPACITANCE IN MFD.	DC RATED VOLTAGE	CASE SIZE	MIL TYPE DESIGNATION
1.8	20	A	CS13AE1R8K
2.2	20	A	AE2R2K
8.2	20	B	AE8R2K
10	20	B	AE100K
12	20	B	AE120K
15	20	B	AE150K
27	20	C	AE270K
33	20	C	AE330K
39	20	C	AE390K
47	20	D	AE470K
56	20	D	AE560K
68	20	D	AE680K
82	20	D	AE820K
100	20	D	AE101K
0.33	35	A	AFR33K
0.39	35	A	AFR39K
0.47	35	A	AFR47K
0.56	35	A	AFR56K
0.68	35	A	AFR68K
0.82	35	A	AFR82K
1	35	A	AF010K
1.2	35	B	AF1R2K
1.5	35	B	AF1R5K
1.8	35	B	AF1R8K
2.2	35	B	AF2R2K
2.7	35	B	AF2R7K
3.3	35	B	AF3R3K
3.9	35	B	AF3R9K
4.7	35	B	AF4R7K
5.6	35	B	CS13AF5R6K

CAPACITANCE IN MFD.	DC RATED VOLTAGE	CASE SIZE	MIL TYPE DESIGNATION
6.8	35	B	CS13AF6R8K
8.2	35	C	AF8R2K
10	35	C	AF100K
12	35	C	AF120K
15	35	C	AF150K
18	35	C	AF180K
22	35	D	AF220K
27	35	D	AF270K
33	35	D	AF330K
39	35	D	AF390K
47	35	D	AF470K
1	50	A	AG010K
1.2	50	B	AG1R2K
1.5	50	B	AG1R5K
1.8	50	B	AG1R8K
2.2	50	B	AG2R2K
2.7	50	B	AG2R7K
3.3	50	B	AG3R3K
3.9	50	B	AG3R9K
4.7	50	B	AG4R7K
5.6	50	C	AG5R6K
6.8	50	C	AG6R8K
8.2	50	C	AG8R2K
10	50	C	AG100K
12	50	C	AG120K
15	50	C	AG150K
18	50	C	AG180K
22	50	D	CS13AG220K

CAPACITOR DIMENSIONS

CASE SIZE	C max	D (0 +.016 −.015)	L ±.031	LEAD DIAMETER (+.005 −.001)
A	0.422	0.135	0.286	0.020
B	0.610	0.185	0.474	0.020
C	0.822	0.289	0.686	0.025
D	0.922	0.351	0.786	0.025

SYMBOL —)|—

COMPONENT OUTLINE (SEE DIMENSIONS)

EXAMPLE:
0.68 MFD, 35V, ±10% = Mil No. CS13AFR68K
For Capacitor Dimensions see Case A

APPENDIX B

DIODE TYPE NUMBER	DIODE TYPE NUMBER	DIODE TYPE NUMBER	DIODE DWG. NO.	MANUFACTURER'S NAME
1N91	1N93		DO-3	GENERAL ELECTRIC
1N92			DO-3	GENERAL ELECTRIC
1N550	1N553		DO-4	GENERAL ELECTRIC
1N551	1N554		DO-4	GENERAL ELECTRIC
1N552	1N555		DO-4	GENERAL ELECTRIC
1N2155	1N2157		DO-5	GENERAL ELECTRIC
1N2154	1N2156		DO-5	GENERAL ELECTRIC
1N3064	1N3600		DO-7	GENERAL ELECTRIC
1N3604	1N3605		DO-7	GENERAL ELECTRIC
1N3606			DO-7	GENERAL ELECTRIC
1N483A	1N617		DO-7	HUGHES
1N1765	1N1776		DO-13	GENERAL ELECTRIC
1N3712	1N3714		A	GENERAL ELECTRIC
1N3713	1N3715		A	GENERAL ELECTRIC

NOTE:
Blank spaces can be filled in by either teacher or student with additional diodes he may need as reference material.

REF. DESIG.	COMPONENT:
CR	RECTIFIER, DIODE

DIODE OUTLINE
DWG. NO.

DWG. NO. DO-3

DWG. NO. DO-4

DIODE SYMBOL

DWG. NO. DO-5

DWG. NO. DO-7

DIODE SYMBOL

DWG. NO. DO-13

DWG. NO. A.

DIODE SYMBOL

COMPONENT OUTLINE

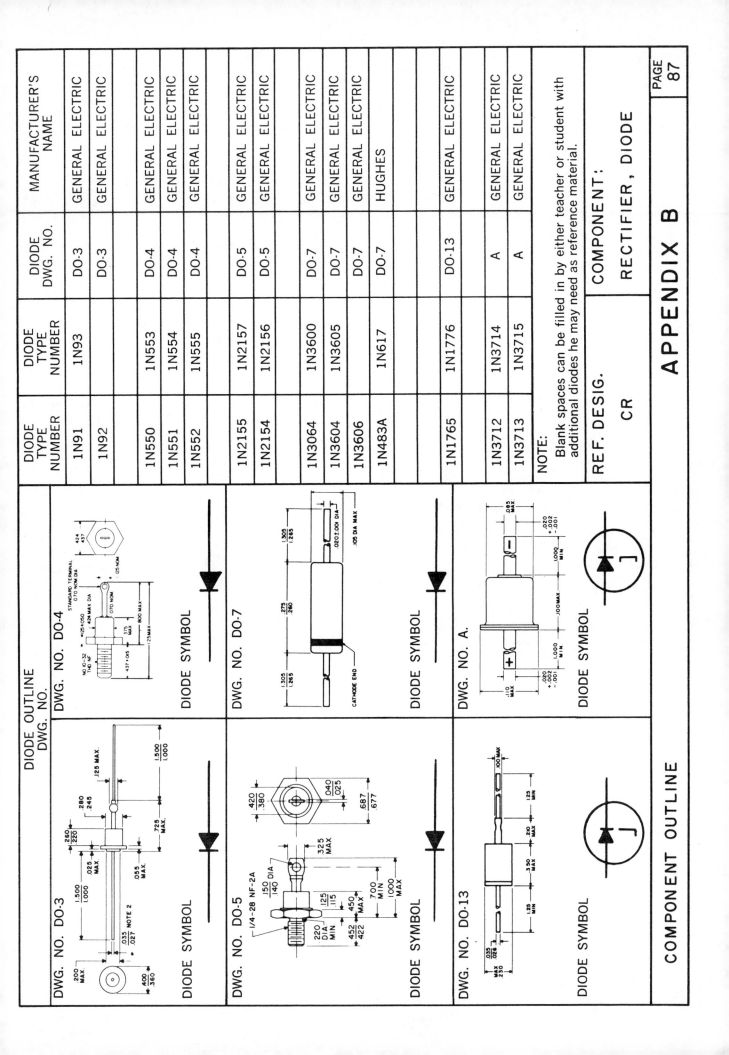

APPENDIX B

APPENDIX B

COMPONENT: RESISTOR, FIXED, COMPOSITION 1/4 WATT ±5% (MIL STYLE RC07)

SYMBOL: ⊸/\/\/\⊸

REF. DESIG.: R

Total Resistance Ohms	MIL Type Designation	Total Resistance Ohms	MIL Type Designation	Total Resistance Ohms	MIL Type Designation	Total Resistance Ohms	MIL Type Designation	Total Resistance Megohms	MIL Type Designation
10	RC07GF100J	180	RC07GF181J	3,300	RC07GF332J	56,000	RC07GF563J	1.0	RC07GF105J
11	110J	200	201J	3,600	362J	62,000	623J	1.1	115J
12	120J	220	221J	3,900	392J	68,000	683J	1.2	125J
13	130J	240	241J	4,300	432J	75,000	753J	1.3	135J
15	150J	270	271J	4,700	472J	82,000	823J	1.5	155J
16	160J	300	301J	5,100	512J	91,000	913J	1.6	165J
18	180J	330	331J	5,600	562J	100,000	104J	1.8	185J
20	200J	360	361J	6,200	622J	110,000	114J	2.0	205J
22	220J	390	391J	6,800	682J	120,000	124J	2.2	225J
24	240J	430	431J	7,500	752J	130,000	134J	2.4	245J
27	270J	470	471J	8,200	822J	150,000	154J	2.7	275J
30	300J	510	511J	9,100	912J	160,000	164J	3.0	305J
33	330J	560	561J	10,000	103J	180,000	184J	3.3	335J
36	360J	620	621J	11,000	113J	200,000	204J	3.6	365J
39	390J	680	681J	12,000	123J	220,000	224J	3.9	395J
43	430J	750	751J	13,000	133J	240,000	244J	4.3	435J
47	470J	820	821J	15,000	153J	270,000	274J	4.7	475J
51	510J	910	911J	16,000	163J	300,000	304J	5.1	515J
56	560J	1,000	102J	18,000	183J	330,000	334J	5.6	565J
62	620J	1,100	112J	20,000	203J	360,000	364J	6.2	625J
68	680J	1,200	122J	22,000	223J	390,000	394J	6.8	685J
75	750J	1,300	132J	24,000	243J	430,000	434J	7.5	755J
82	820J	1,500	152J	27,000	273J	470,000	474J	8.2	825J
91	910J	1,600	162J	30,000	303J	510,000	514J	9.1	915J
100	101J	1,800	182J	33,000	333J	560,000	564J	10.0	106J
110	111J	2,000	202J	36,000	363J	620,000	624J	11	116J
120	121J	2,200	222J	39,000	393J	680,000	684J	12	126J
130	131J	2,400	242J	43,000	433J	750,000	754J	13	136J
150	151J	2,700	272J	47,000	473J	820,000	824J	15	156J
160	RC07GF161J	3,000	RC07GF302J	51,000	RC07GF513J	910,000	RC07GF914J	16	166J
								18	186J
								20	206J
								22	RC07GF226J

COMPONENT OUTLINE

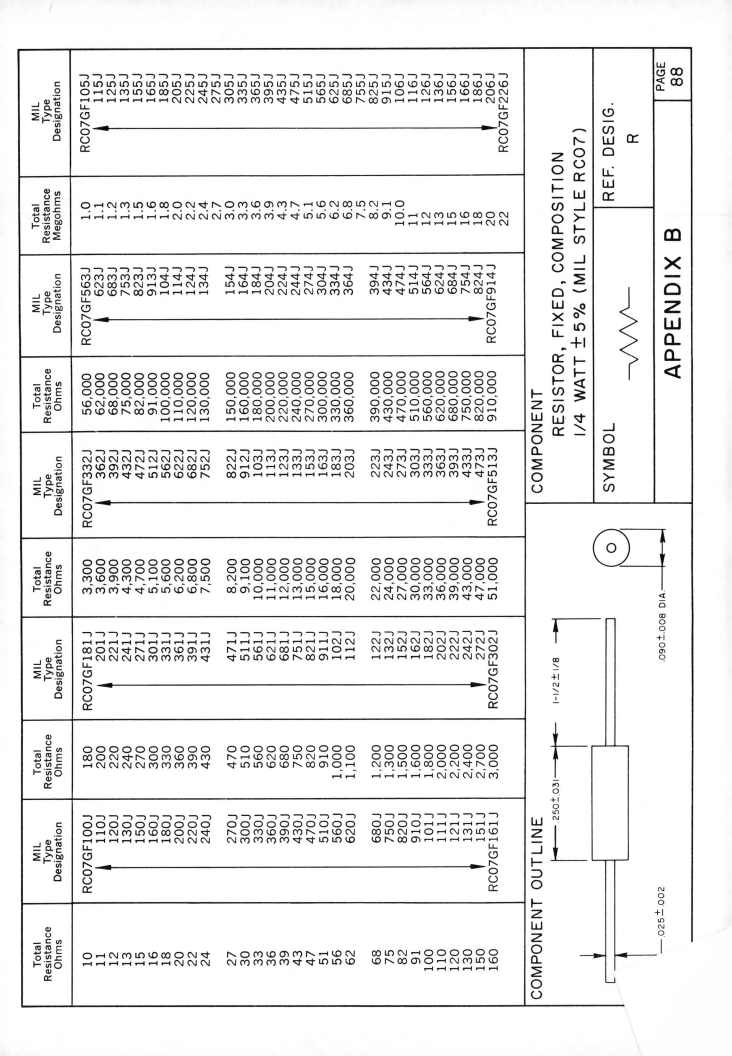

.250 ± .031
1-1/2 ± 1/8
.090 ± .008 DIA.
.025 ± .002

APPENDIX B

COMPONENT: RESISTOR, FIXED, COMPOSITION 1/2 WATT ± 5% (MIL STYLE RC20)

SYMBOL: ⎓ (resistor symbol) **REF. DESIG.:** R

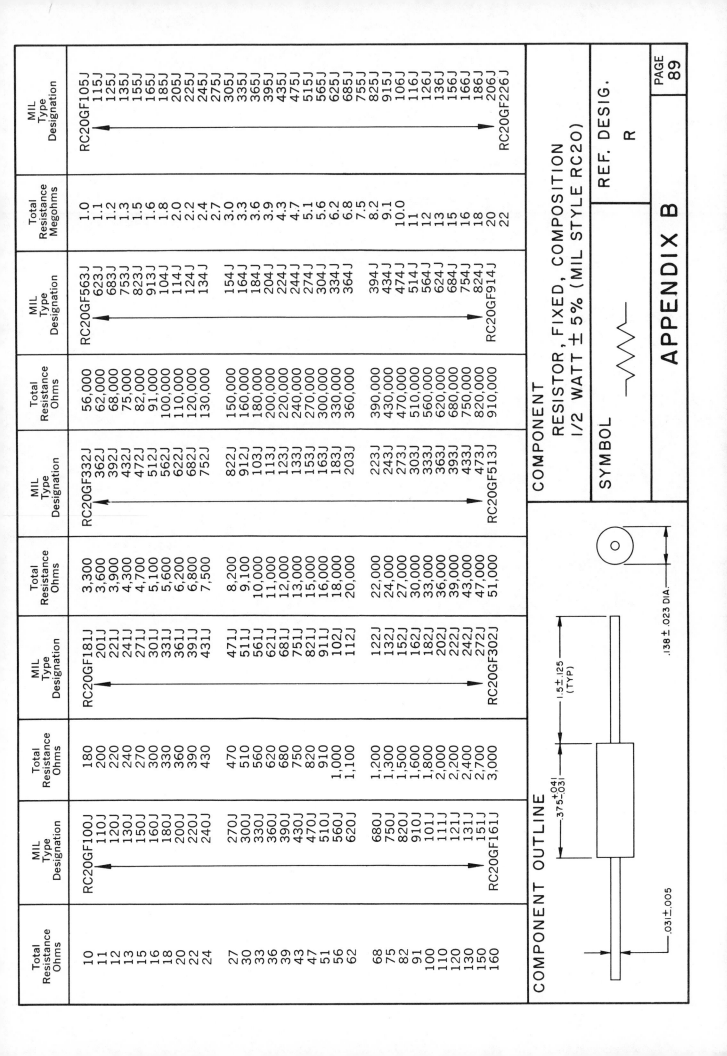

Total Resistance Ohms	MIL Type Designation	Total Resistance Ohms	MIL Type Designation	Total Resistance Ohms	MIL Type Designation	Total Resistance Ohms	MIL Type Designation	Total Resistance Megohms	MIL Type Designation
10	RC20GF100J	180	RC20GF181J	3,300	RC20GF332J	56,000	RC20GF563J	1.0	RC20GF105J
11	110J	200	201J	3,600	362J	62,000	623J	1.1	115J
12	120J	220	221J	3,900	392J	68,000	683J	1.2	125J
13	130J	240	241J	4,300	432J	75,000	753J	1.3	135J
15	150J	270	271J	4,700	472J	82,000	823J	1.5	155J
16	160J	300	301J	5,100	512J	91,000	913J	1.6	165J
18	180J	330	331J	5,600	562J	100,000	104J	1.8	185J
20	200J	360	361J	6,200	622J	110,000	114J	2.0	205J
22	220J	390	391J	6,800	682J	120,000	124J	2.2	225J
24	240J	430	431J	7,500	752J	130,000	134J	2.4	245J
								2.7	275J
27	270J	470	471J	8,200	822J	150,000	154J	3.0	305J
30	300J	510	511J	9,100	912J	160,000	164J	3.3	335J
33	330J	560	561J	10,000	103J	180,000	184J	3.6	365J
36	360J	620	621J	11,000	113J	200,000	204J	3.9	395J
39	390J	680	681J	12,000	123J	220,000	224J	4.3	435J
43	430J	750	751J	13,000	133J	240,000	244J	4.7	475J
47	470J	820	821J	15,000	153J	270,000	274J	5.1	515J
51	510J	910	911J	16,000	163J	300,000	304J	5.6	565J
56	560J	1,000	102J	18,000	183J	330,000	334J	6.2	625J
62	620J	1,100	112J	20,000	203J	360,000	364J	6.8	685J
								7.5	755J
68	680J	1,200	122J	22,000	223J	390,000	394J	8.2	825J
75	750J	1,300	132J	24,000	243J	430,000	434J	9.1	915J
82	820J	1,500	152J	27,000	273J	470,000	474J	10.0	106J
91	910J	1,600	162J	30,000	303J	510,000	514J	11	116J
100	101J	1,800	182J	33,000	333J	560,000	564J	12	126J
110	111J	2,000	202J	36,000	363J	620,000	624J	13	136J
120	121J	2,200	222J	39,000	393J	680,000	684J	15	156J
130	131J	2,400	242J	43,000	433J	750,000	754J	16	166J
150	151J	2,700	272J	47,000	473J	820,000	824J	18	186J
160	RC20GF161J	3,000	RC20GF302J	51,000	RC20GF513J	910,000	RC20GF914J	20	206J
								22	RC20GF226J

COMPONENT OUTLINE

.031 ± .005
.375 +.041 −.031
1.5 ± .125 (TYP)
.138 ± .023 DIA.

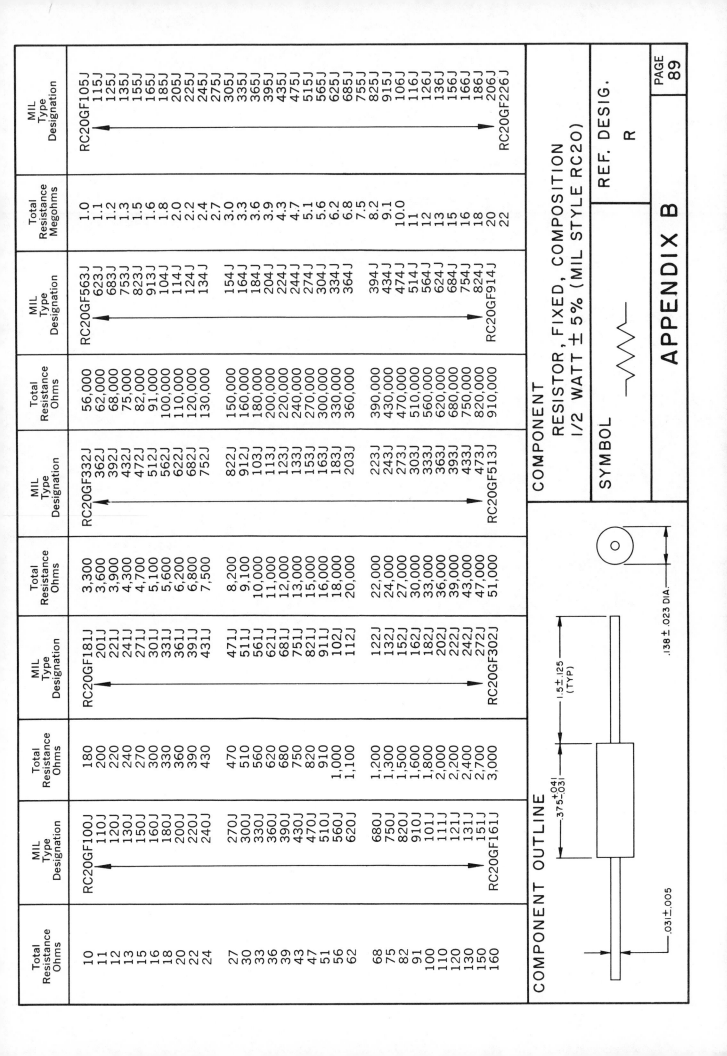

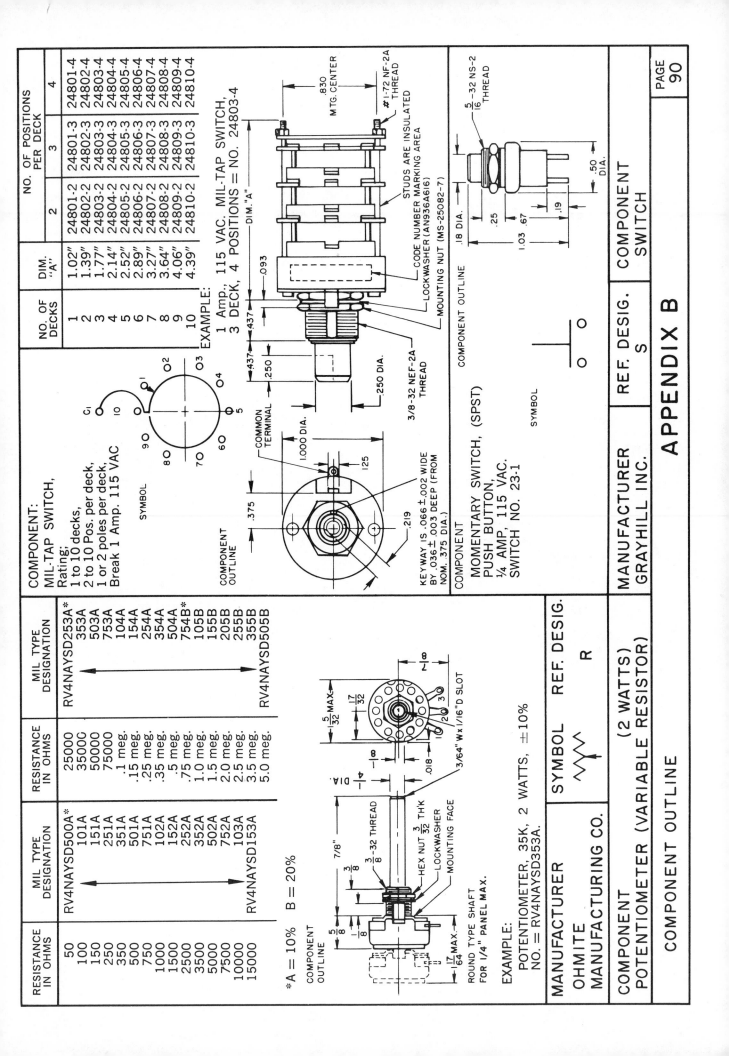

APPENDIX B

COMPONENT OUTLINE

Potentiometer (Variable Resistor)

RESISTANCE IN OHMS	MIL TYPE DESIGNATION	RESISTANCE IN OHMS	MIL TYPE DESIGNATION
50	RV4NAYSD500A*	25000	RV4NAYSD253A*
100	101A	35000	353A
150	151A	50000	503A
250	251A	75000	753A
350	351A	.1 meg.	104A
500	501A	.15 meg.	154A
750	751A	.25 meg.	254A
1000	102A	.35 meg.	354A
1500	152A	.5 meg.	504A
2500	252A	.75 meg.	754B*
3500	352A	1.0 meg.	105B
5000	502A	1.5 meg.	155B
7500	752A	2.0 meg.	205B
10000	103A	2.5 meg.	255B
15000	RV4NAYSD153A	3.5 meg.	355B
		5.0 meg.	RV4NAYSD505B

*A = 10%   B = 20%

COMPONENT OUTLINE

EXAMPLE:
POTENTIOMETER, 35K, 2 WATTS, ±10%
NO. = RV4NAYSD353A.

MANUFACTURER	SYMBOL	REF. DESIG.
OHMITE MANUFACTURING CO.	⌇	R

COMPONENT
POTENTIOMETER (VARIABLE RESISTOR) (2 WATTS)

ROUND TYPE SHAFT FOR 1/4" PANEL MAX.

3/8-32 THREAD
HEX NUT 3/32 TH'K
LOCKWASHER
MOUNTING FACE
7/8"
3/8
5/8
1/8
1/4 DIA.
17/64 MAX.
5/32 MAX.
17/32
8/7
8/1
.018
3/64" W x 1/16" D SLOT

MIL-TAP Switch

NO. OF DECKS	DIM. "A"	NO. OF POSITIONS PER DECK		
		2	3	4
1	1.02"	24801-2	24801-3	24801-4
2	1.39"	24802-2	24802-3	24802-4
3	1.77"	24803-2	24803-3	24803-4
4	2.14"	24804-2	24804-3	24804-4
5	2.52"	24805-2	24805-3	24805-4
6	2.89"	24806-2	24806-3	24806-4
7	3.27"	24807-2	24807-3	24807-4
8	3.64"	24808-2	24808-3	24808-4
9	4.06"	24809-2	24809-3	24809-4
10	4.39"	24810-2	24810-3	24810-4

COMPONENT:
MIL-TAP SWITCH,
Rating;
1 to 10 decks,
2 to 10 Pos. per deck,
1 or 2 poles per deck,
Break 1 Amp. 115 VAC

SYMBOL
C₁, I₀, ₀1, ₀2, ₀3, ₀4, ₀5, 6₀, 7₀, 8₀, 9₀

EXAMPLE:
1 Amp., 115 VAC. MIL-TAP SWITCH,
3 DECK, 4 POSITIONS = NO. 24803-4

#1-72 NF-2A THREAD
.830 MTG. CENTER
STUDS ARE INSULATED
CODE NUMBER MARKING AREA
LOCKWASHER (AN936A6I6)
MOUNTING NUT (MS-25082-7)
DIM. "A"
.093
.437
.250
.250 DIA.
3/8-32 NEF-2A THREAD
COMMON TERMINAL
1.000 DIA.
.125
.375
.219
COMPONENT OUTLINE
KEYWAY IS .066 ±.002 WIDE BY .036 ±.003 DEEP (FROM NOM. .375 DIA.)

Momentary Switch

MANUFACTURER	REF. DESIG.	COMPONENT
GRAYHILL INC.	S	SWITCH

COMPONENT
MOMENTARY SWITCH, (SPST)
PUSH BUTTON,
1/4 AMP, 115 VAC.
SWITCH NO. 23-1

SYMBOL

5/16-32 NS-2 THREAD
.50 DIA.
.18 DIA.
.25
.67
.19
1.03
COMPONENT OUTLINE

37.4

APPENDIX B

TRANSISTOR SYMBOL	MANUFACTURER'S NAME	TRANS. DWG. NO.	TRANSISTOR TYPE NUMBER	TRANSISTOR TYPE NUMBER
NPN	GENERAL ELECTRIC	TO-5	2N337	2N1613
	GENERAL ELECTRIC	TO-5	2N338	2N1711
	TEXAS INSTRUMENTS	TO-5	2N1304	
PNP	GENERAL ELECTRIC	TO-5	2N427	2N428
	GENERAL ELECTRIC	TO-5	2N395	2N396
	GENERAL ELECTRIC	TO-5	2N404	
NPN	TEXAS INSTRUMENTS	TO-5	2N1305	
	GENERAL ELECTRIC	TO-18	2N914	2N871
	HUGHES	TO-18	2N910	2N911
NPN	GENERAL ELECTRIC	A	2N1047	2N1049
	GENERAL ELECTRIC	A	2N1048	2N1050
NPN	GENERAL ELECTRIC	B	2N2417A	2N2646
	GENERAL ELECTRIC	B	2N2418A	2N2647
	GENERAL ELECTRIC	B	2N2419A	

NOTE:
Blank spaces can be filled in by either teacher or student with additional transistors he may need as reference material.

REF. DESIG.	COMPONENT
Q	TRANSISTOR

TRANSISTOR OUTLINE — DWG. NO.

DWG. NO. TO-18

DWG. NO. TO-5

DWG. NO. A

DWG. NO. B

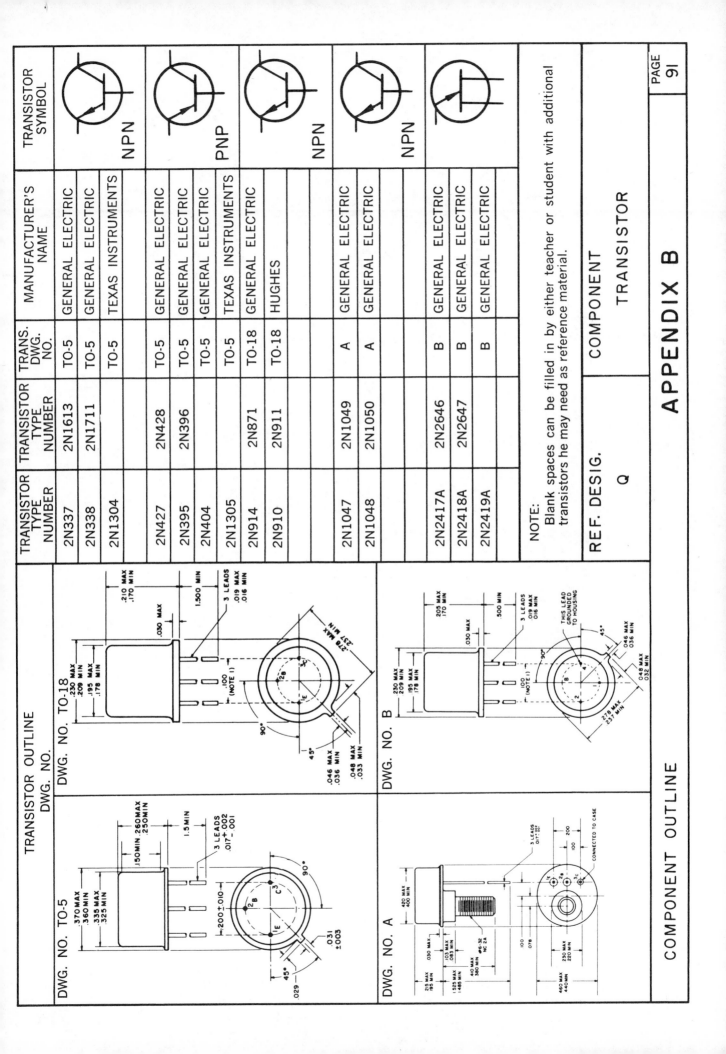

COMPONENT OUTLINE

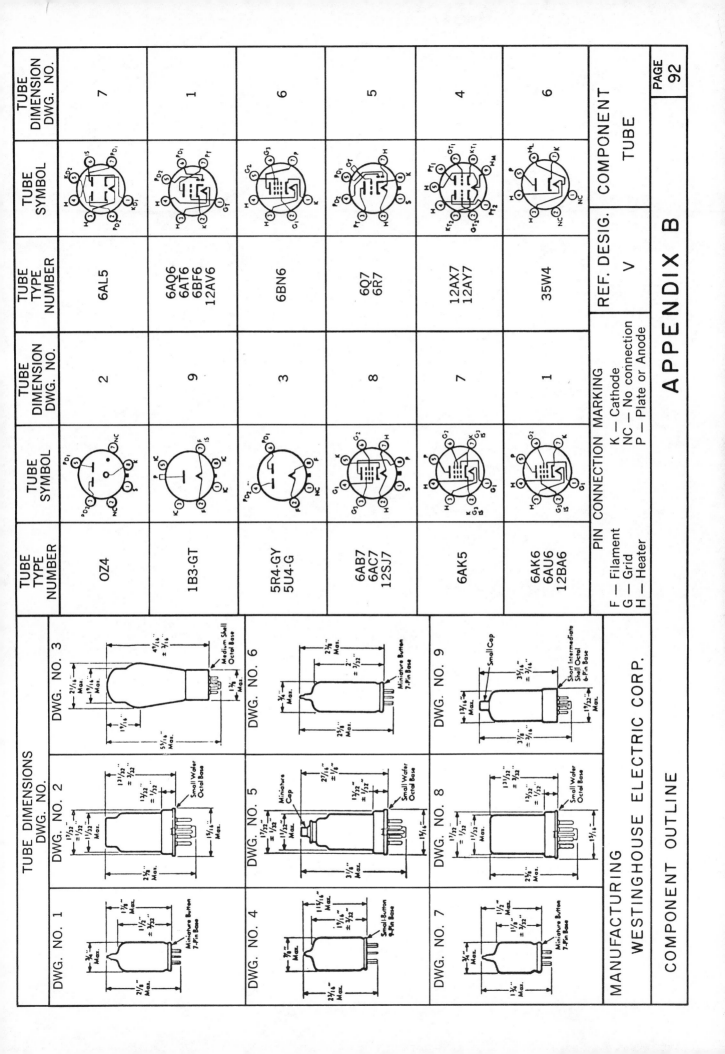

APPENDIX B

COMPONENT OUTLINE

MANUFACTURING — WESTINGHOUSE ELECTRIC CORP.

TUBE TYPE NUMBER	TUBE SYMBOL	TUBE DIMENSION DWG. NO.	TUBE TYPE NUMBER	TUBE SYMBOL	TUBE DIMENSION DWG. NO.
OZ4		2	6AL5		7
1B3-GT		9	6AQ6 / 6AT6 / 6BF6 / 12AV6		1
5R4-GY / 5U4-G		3	6BN6		6
6AB7 / 6AC7 / 12SJ7		8	6Q7 / 6R7		5
6AK5		7	12AX7 / 12AY7		4
6AK6 / 6AU6 / 12BA6		1	35W4		6

PIN CONNECTION MARKING

F — Filament
G — Grid
H — Heater
K — Cathode
NC — No connection
P — Plate or Anode

REF. DESIG.	COMPONENT
V	TUBE

TUBE DIMENSIONS DWG. NO.

DWG. NO. 1 — Miniature Button 7-Pin Base
DWG. NO. 2 — Small Wafer Octal Base
DWG. NO. 3 — Medium Shell Octal Base
DWG. NO. 4 — Small-Button 9-Pin Base
DWG. NO. 5 — Miniature Cap / Small Wafer Octal Base
DWG. NO. 6 — Miniature Button 7-Pin Base
DWG. NO. 7 — Miniature Button 7-Pin Base
DWG. NO. 8 — Small Wafer Octal Base
DWG. NO. 9 — Small Cap / Short Intermediate Shell Octal 6-Pin Base

COMPONENT:
AUDIO OUTPUT TRANSFORMER

TYPE NO.	PRIMARY IMPEDANCE	D.C. Ma	AUDIO WATT
S-10X	10000	45	4-6
S-20X	2000	50	2-3
S-40X	14000	5.5	1/4

CASE DIMENSIONS

	H	W	D	MW
	1 11/16	2 1/16	1 1/2	2 3/8
	1 3/16	2 1/8	1 1/4	1 3/4
	13/16	1 5/8	7/8	1 3/8

EXAMPLE; OUTPUT TRANSFORMER,
Primary Impedance — 2000
50 Ma D.C., 2-3 Watt,
NO. of Transformer — S-20X

SYMBOL

COMPONENT OUTLINE

COMPONENT;
LOW VOLTAGE TRANSFORMER, 1 AMP. D.C.
Transformer — F-92A

CASE DIMENSIONS

H — 3½   MW — 2
W — 2 23/32   MD — 2¼
D — 3
T — 1.0   TD — 2¼

.180 DIA.
4 HOLES
.438 DIA.
1/16

SYMBOL
G (GREEN)
R (RED)
G/B
B/R

COMPONENT OUTLINE

MANUFACTURER	REF. DESIG.	COMPONENT
TRIAD TRANSFORMER CORPORATION	T	TRANSFORMER

COMPONENT:
RECEIVING CRYSTAL

FREQUENCY MEGACYCLES	CRYSTAL NO.
26.510	3647
26.570	3652
26.750	3666

SYMBOL

COMPONENT OUTLINE:
.150
.040
.170
.510
.400
.75
.192

MANUFACTURER	REF. DESIG.	COMPONENT
HERMAN H. SMITH, INC.	Y	CRYSTAL

COMPONENT:
FUSE HOLDER FOR 3AG. HOLDER NO. 342001

COMPONENT OUTLINE:

ACROSS FLATS ON THREAD
BACK VIEW
.446
HOLE, WIRE LEAD 3/32
.813
.435
2 7/32
45/64
DIA .709/.722

SYMBOL

COMPONENT:
3 AG QUICK ACTING FUSE,
1/8 Amp, 250V NO. 312.125
3 AG "SLO-BLO" FUSE,
1/2 Amp., 125V NO. 313.500
1 Amp, 125V NO. 313001

COMPONENT OUTLINE:
1¼
1/4 DIA.

MANUFACTURER	REF. DESIG.	COMPONENT
LITTELFUSE, INC.	F	FUSE AND FUSE HOLDER

COMPONENT OUTLINE

APPENDIX B

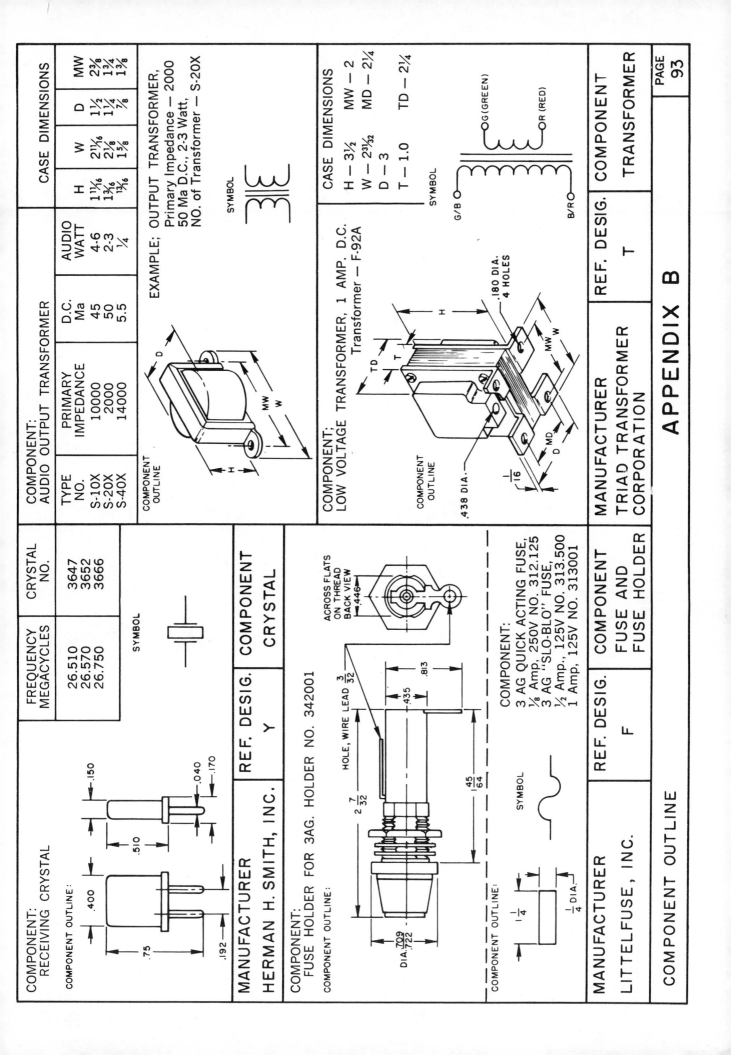

COMPONENT OUTLINE

1 3/8 DIA.

3.56

9/32

5/8

1 15/32 DIA.

RECOMMENDED CHASSIS
CUT OUT

.22

.875
DIA.

.065

.500

SYMBOL

COMPONENT NO. CTM-1284

MANUFACTURER ARCO ELECTRONIC INC.	REF. DESIG. C	COMPONENT: 1500 MFD, 50V ELECTROLYTIC CAPACITOR

COMPONENT OUTLINE

.50

.25

1.50

1.125

.750

.375

.19

.191 DIA.
2 HOLES

.093

.906

.12 DIA.
(TYP.)

1/16 x 3/32
SLOT

SYMBOL

COMPONENT OUTLINE

MANUFACTURER HERMAN H. SMITH, INC.	REF. DESIG. J	COMPONENT: JACK, 2 PINS	COMPONENT NO. 1982

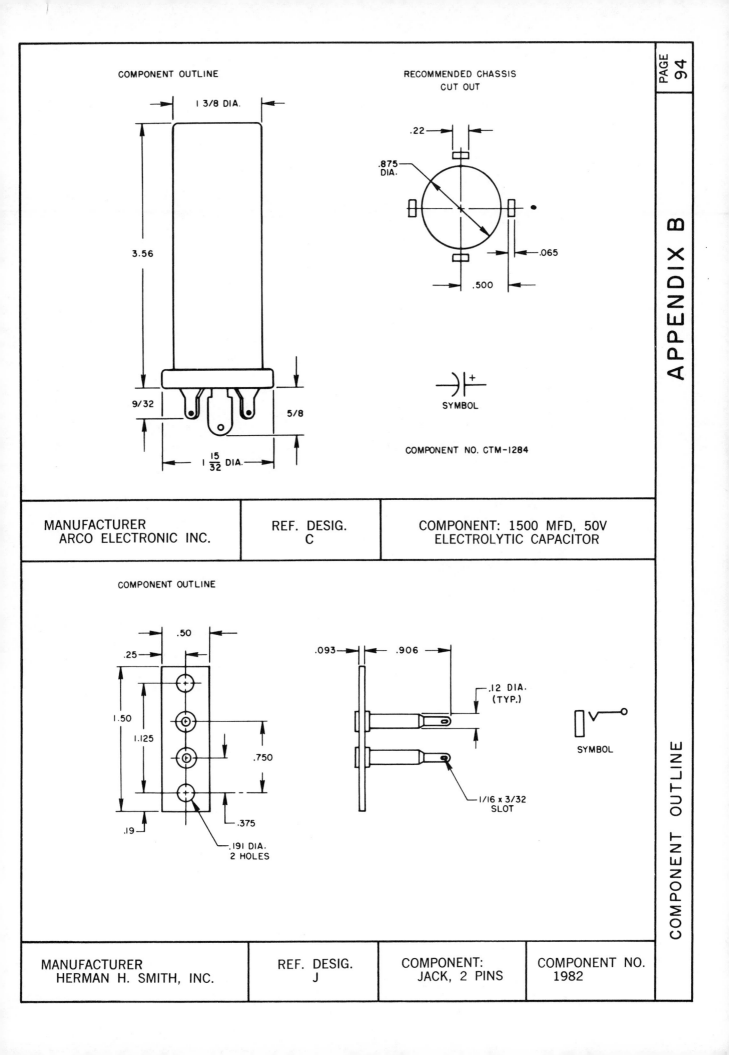

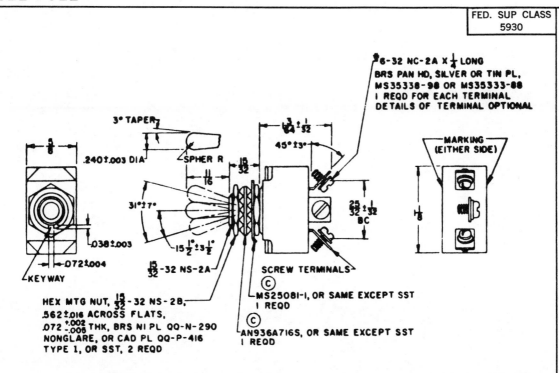

FED. SUP CLASS
5930

MS PART NO.	FORMER MS PART NO.		FORMER AN PART NO.	FORMER JAN TYPE DESIGNATION	CIRCUIT WITH TOGGLE LEVER IN		
	SCREW-LUG TERMINAL	SOLDER[1]/-LUG TERMINAL			Up position	Center position	Down position (keyway side)
MS35058-21	MS35058-5	—	AN3021-1	ST40E	On	Off	On
	—	MS35058-16	—	ST42E			
MS35058-22	—	—	AN3021-2	ST40A	On	None	Off
	—	MS35058-9	—	ST42A			
MS35058-23	MS35058-4	—	AN3021-3	ST40D	On	None	On
	—	MS35058-15	—	ST42D			
MS35058-24	MS35058-1	—	AN3021-10	—	On	Off	None
	—	MS35058-12	—	—			
MS35058-25	MS35058-3	—	AN3021-12	—	On	Mom off	None
	—	MS35058-14	—	—			
MS35038-26	MS35058-6	—	—	ST40F	On	None	Mom on
	—	MS35058-17	—	ST42F			
MS35058-27	MS35058-7	—	AN3021-7	ST40G	Mom on	Off	Mom on
	—	MS35058-18	—	ST42G			
MS35058-28	MS35058-2	—	AN3021-11	—	None	Off	Mom on
	—	MS35058-13	—	—			
MS35058-29	—	—	AN3021-9	ST40B	On	None	Mom off
	—	MS35058-10	—	ST42B			
MS35058-30	—	—	AN3021-8	ST40C	Off	None	Mom on
	—	MS35058-11	—	ST42C			
MS35058-31	MS35058-8	—	AN3021-6[2]	ST40H	On	Off	Mom on
	—	MS35058-19	—	ST42H			

[1]These MS part numbers are superseded by MS part numbers MS35058-21 to MS35058-31, inclusive, as applicable. By removing the screws and lockwashers from the applicable part, a switch with solder-lug terminals is formed.
[2]With jumper removed.

APPROVED 24 July 1951   REVISED Ⓐ 17 May 1955 Ⓑ 15 April 1958 © 20 June 1960

P.A. USAF	TITLE	MILITARY STANDARD
Other Cust SigC Wep	SWITCHES, TOGGLE, SINGLE POLE, ONE-HOLE MOUNTING (SEALED TOGGLE BUSHING)	
		MS35058
PROCUREMENT SPECIFICATION MIL-S-3950	SUPERSEDES: Air Force — Navy Aeronautical Standard AN3021 dated 12 January 1954	SHEET 1 OF

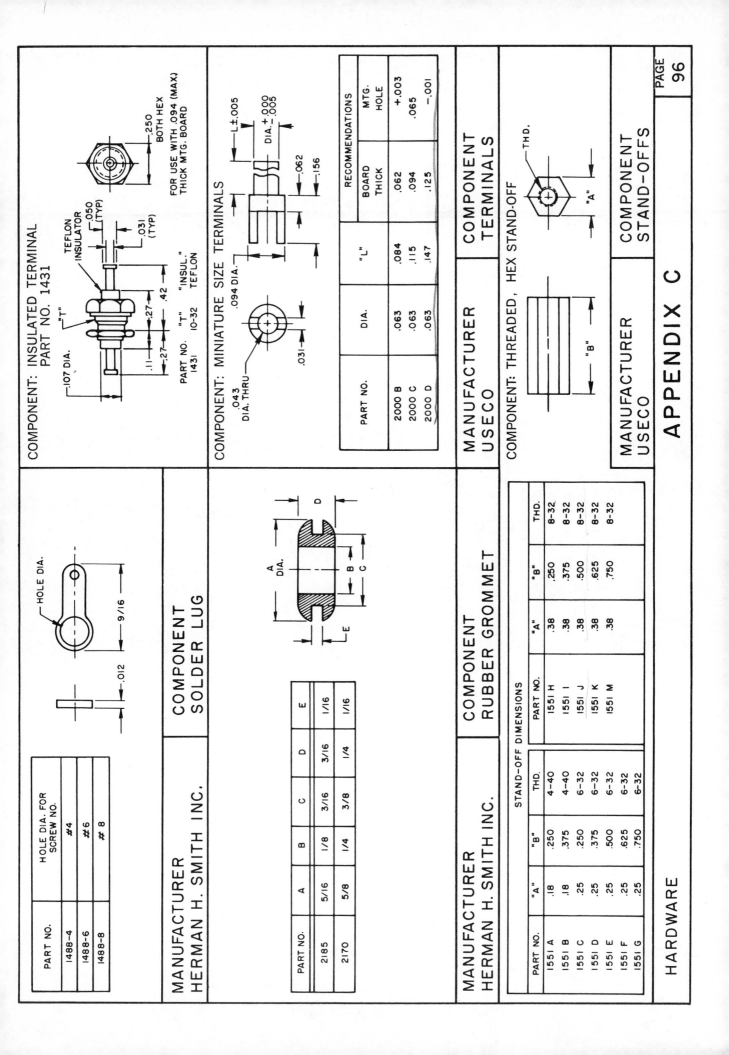

COMPONENT: INSULATED TERMINAL
PART NO. 1431

PART NO.	"T"
1431	10-32

.250 BOTH HEX (MAX)
FOR USE WITH .094 (MAX) THICK MTG. BOARD

MANUFACTURER USECO — COMPONENT TERMINALS

COMPONENT: MINIATURE SIZE TERMINALS

PART NO.	DIA.	"L"
2000 B	.063	.084
2000 C	.063	.115
2000 D	.063	.147

RECOMMENDATIONS

BOARD THICK	MTG. HOLE
.062	.065 +.003 -.001
.094	
.125	

COMPONENT: THREADED, HEX STAND-OFF

MANUFACTURER USECO — COMPONENT STAND-OFFS

APPENDIX C

PART NO.	HOLE DIA. FOR SCREW NO.
1488-4	#4
1488-6	#6
1488-8	#8

MANUFACTURER HERMAN H. SMITH INC. — COMPONENT SOLDER LUG

PART NO.	A	B	C	D	E
2185	5/16	1/8	3/16	3/16	1/16
2170	5/8	1/4	3/8	1/4	1/16

MANUFACTURER HERMAN H. SMITH INC. — COMPONENT RUBBER GROMMET

STAND-OFF DIMENSIONS

PART NO.	"A"	"B"	THD.	PART NO.	"A"	"B"	THD.
1551 A	.18	.250	4-40	1551 H	.38	.250	8-32
1551 B	.18	.375	4-40	1551 I	.38	.375	8-32
1551 C	.25	.250	6-32	1551 J	.38	.500	8-32
1551 D	.25	.375	6-32	1551 K	.38	.625	8-32
1551 E	.25	.500	6-32	1551 M	.38	.750	8-32
1551 F	.25	.625	6-32				
1551 G	.25	.750	6-32				

HARDWARE

FED SUP CLASS
5305

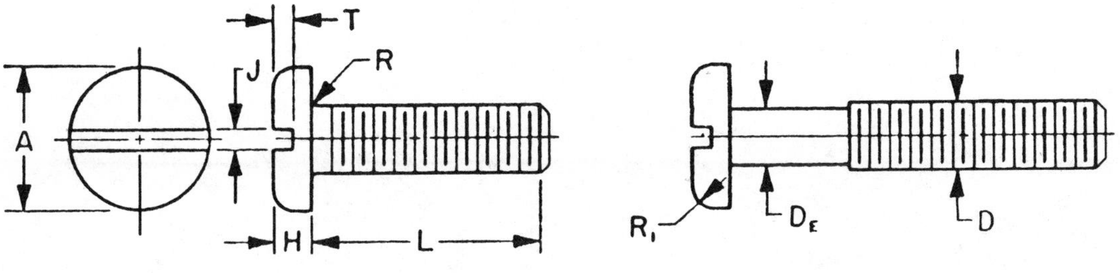

Nominal Size	D		2 (.086)	4 (.112)	6 (.138)	8 (.164)
Threads Per Inch			56NC	40NC	32NC	32NC
Body Diameter	D_E	Max Min	.0860 .0717	.1120 .0925	.1380 .1141	.1640 .1399
Head Diameter	A	Max Min	.167 .155	.219 .205	.270 .256	.322 .306
Head Height	H	Max Min	.053 .045	.068 .058	.082 .072	.096 .085
Slot Width	J	Max Min	.031 .023	.039 .031	.048 .039	.054 .045
Slot Depth	T	Max Min	.033 .023	.041 .030	.050 .038	.058 .043
Radius On Head	R_1	Nom	.035	.042	.046	.052
Radius Under Head	R	Max	.013	.018	.023	.023

Length	L	Tolerance	Dash No.	FIIN	Dash No.	FIIN	Dash No.	FIIN	Dash No.	FIIN
Threads shall extend to within 2 threads of the bearing surface of the head, or closer if practicable.	1/8		1		11	043-6472	24	043-6500	39	043-6528
	3/16		2		12	043-6473	25	019-3254	40	043-6529
	1/4		3		13	043-6474	26	019-3256	41	043-6530
	5/16		4		14	043-6475	27	043-6501	42	043-6531
	3/8	+0	5		15	043-6476	28	043-6502	43	043-6532
	7/16	−1/32	6		16	043-6477	29	043-6503	44	043-6533
	1/2		7		17	043-6478	30	043-6504	45	043-6534
	5/8		8		18	043-6479	31	043-6505	46	043-6535
	3/4		9		19	043-6480	32	043-6506	47	043-6536
	7/8		10		20	043-6481	33	043-6507	48	043-6537
	1				21	043-6482	34	043-6508	49	043-6538
	1 1/4	+0			22	043-6483	35	043-6509	50	043-6539
	1 1/2	−1/16			23	043-6486	36	043-6510	51	043-6540
	1 3/4						37	043-6511	52	043-6541
	2						38	043-6512	53	043-6542
Minimum complete thread length of 1 3/4.	2 1/4								54	043-6543
	2 1/2	+0							55	043-6544
	2 3/4	−3/32							56	043-6545
	3								57	043-6546

Materials: Carbon Steel, Specification QQ-W-409 or QQ-S-633, except Bessemer Compositions; 55,000 PSI minimum ultimate tensile strength.

Finish: Plain (untreated).

Threads: The threads shall be in accordance with Screw-thread Standards for Federal Services Handbook H-28.

Notes:
(1) Referenced documents of issue in effect on date of invitation for bids shall apply.
(2) In case of conflict with any referenced document, this standard will govern.
(3) The MS part number consists of the MS sheet number, plus the dash number. Example: MS 35221-1.
(4) All dimensions in inches.

REVISED

APPROVED FEB 9 1954

CUSTODIANS	OTHER INT.	MILITARY STANDARD	MS35221
A — ORD N — SHIPS AF — USAF	A — CEMQSigT N — AMdOrSYMC AF —	SCREW, MACHINE, PAN HEAD, SLOTTED, CARBON STEEL, PLAIN FINISH, NC-2A AND UNC-2A	
PROCUREMENT SPECIFICATION FF — S — 92		SUPERSEDES:	SHEET 1 OF 2

ACCEPTABLE DESIGNS

A	Nominal Size or Basic Major Dia. of Thread		No. 2 .086		No. 4 .112		No. 6 .138		No. 8 .164	
B	Threads per Inch		56		40		32		32	
C	Width Across Flats	Nom Max Min	3/16 .1875 .180		1/4 .2500 .241		5/16 .3125 .302		11/32 .3438 .332	
D	Width Across Corners	Max Min	.217 .205		.289 .275		.361 .344		.397 .378	
E	Thickness	Nom Max Min	1/16 .066 .057		3/32 .098 .087		7/64 .114 .102		1/8 .130 .117	
	Material and Protective Coating	Dash No.	FIIN	Dash No.	FIIN	Dash No.	FIIN	Dash No.	FIIN	
	Steel, Carbon Uncoated Cadmium or Zinc Optional Phosphate	21 22 23	019-1716	41 42 43	011-4776 013-4524 275-9301	61 62 63	275-1706 013-4530	81 82 83	275-6800 012-0622 275-9310	
	Steel, Corrosion Resisting Passivated	24	271-4640	44	271-4642	64	271-4644	84	271-4645	
	Brass Uncoated Tin Plated	25 26		45 46		65 66		85 86		

A	Nominal Size or Basic Major Dia. of Thread		No. 10 .190							
B	Threads per Inch		24							
C	Width Across Flats	Nom Max Min	3/8 .3750 .362							
D	Width Across Corners	Max Min	.433 .413							
E	Thickness	Nom Max Min	1/8 .130 .117							
	Material and Protective Coating	Dash No.	FIIN	Dash No.	FIIN	Dash No.	FIIN	Dash No.	FIIN	
	Steel, Carbon Uncoated Cadmium or Zinc Optional Phosphate	101 102 103	350-3384 012-0361 281-5341							
	Steel, Corrosion Resisting Passivated	104	275-5095							
	Brass Uncoated Tin Plated	105 106								

Material: Steel, Carbon, (Commercial Grade) except Bessemer Steels.
Steel, Corrosion Resisting, Federal Standard No. 66, Steel Numbers: #303, 304, 305, 410, 416, 430.
Brass, Naval, Specification MIL-B-994, Composition A or C.

Protective Coating: Cadmium Plate, Specification QQ-P-416, Type II, Class C.
Zinc Plate, Specification QQ-Z-325, Type II, Class 3.
Phosphate, Specification MIL-C-16232, Type II.
Tin Plate, Specification MIL-T-10727, Type I or II, .0001 thick.

Thread: The threads shall be in accordance with Screw-Thread Standards for Federal Services, Handbook H-28.

Notes:
(1) Referenced documents shall be of the issue in effect on date of invitations for bid.
(2) This document has been promulgated by the Department of Defense as the Military Standard to limit the selection of the item, product or design covered herein in engineering, design and procurement. This standard shall become effective not later than 90 days after the latest date of approval shown.
(3) This standard takes precedence over documents referenced herein.
(4) Nuts shall be free from burrs, scale and all other defects that would affect their serviceability.
(5) The MS part number consists of the MS sheet number, plus the dash number. Example: MS35649-22.
(6) All dimensions in inches.

APPROVED 22 DEC 55   REVISED

CUSTODIANS	OTHER INT.	MILITARY STANDARD	
A — ORD N — SHIPS AF — USAF	A — CESiGT N — MCORY AF —	NUT, PLAIN, HEXAGON, MACHINE SCREW, NC-2B	MS35649
PROCUREMENT SPECIFICATION FF-N-836		SUPERSEDES:	SHEET 1 OF 1

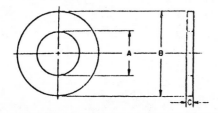

TOLERANCES

±.010 ON OUTSIDE DIAMETER.
±.005 ON INSIDE DIAMETER UP TO AND INCLUDING #10.
±.010 ON ALL OTHER INSIDE DIAMETERS.
  INSIDE AND OUTSIDE DIAMETERS SHALL BE CONCENTRIC
  WITHIN THE TOLERANCE OF THE INSIDE DIAMETER.

NOM. SIZE	"A" I.D.	"B" O.D.	"C" THICKNESS Max.	Min.	CRES Dash No.	FIIN	Ni-Cu ALLOY Dash No.	FIIN	COPPER Dash No.	FIIN	BRASS Dash No.	FIIN	ALUMINUM ALLOY Dash No.	FIIN
0	.078	.187	.025	.016	301		401		501		601		701	
2	.093	.250	.025	.016	302		402		502		602		702	
4	.125	.250	.028	.017	303		403		503		603		703	
4	.125	.312	.040	.025	304		404		504		604		704	
6	.156	.312	.048	.027	305		405		505		605		705	
6	.156	.375	.065	.036	306		406		506		606		706	
8	.187	.375	.065	.036	307		407		507		607		707	
10	.218	.437	.065	.036	308		408		508		608		708	
10	.250	.562	.080	.051	309		409		509		609		709	
¼	.281	.625	.080	.051	310		410		510		610		710	
¼	.312	.750	.080	.051	311		411		511		611		711	
⁵⁄₁₆	.343	.687	.080	.051	312		412		512		612		712	
⁵⁄₁₆	.375	.875	.104	.064	313		413		513		613		713	
⅜	.406	.812	.080	.051	314		414		514		614		714	
⅜	.437	1.000	.104	.064	315		415		515		615		715	
⁷⁄₁₆	.468	.921	.080	.051	316		416		516		616		716	
⁷⁄₁₆	.500	1.250	.104	.064	317		417		517		617		717	
½	.531	1.062	.121	.074	318		418		518		618		718	
½	.562	1.375	.132	.086	319		419		519		619		719	
⅝	.656	1.312	.121	.074	320		420		520		620		720	
⅝	.687	1.750	.160	.108	321		421		521		621		721	
¾	.812	1.500	.160	.108	322		422		522		622		722	
¾	.812	2.000	.177	.122	323		423		523		623		723	
⅞	.937	1.750	.160	.108	324		424		524		624		724	
⅞	.937	2.250	.192	.136	325		425		525		625		725	
1	1.062	2.000	.160	.108	326		426		526		626		726	
1	1.062	2.500	.192	.136	327		427		427		627		727	
1⅛	1.250	2.750	.192	.136	328		428		528		628		728	
1¼	1.375	3.000	.192	.136	329		429		529		629		729	
1⅜	1.500	3.250	.213	.153	330		430		530		630		730	
1½	1.625	3.500	.213	.153	331		431		531		631		731	
1⅝	1.750	3.750	.213	.153	332		432		532		632		732	
1¾	1.875	4.000	.213	.153	333		433		533		633		733	
1⅞	2.000	4.250	.213	.153	334		434		534		634		734	
2	2.125	4.500	.213	.153	335		435		535		635		735	
2¼	2.375	4.750	.248	.193	336		436		536		636		736	
2½	2.625	5.000	.280	.210	337		437		537		637		737	
2¾	2.875	5.250	.310	.228	338		438		538		638		738	
3	3.125	5.500	.327	.249	339		439		539		639		739	

All dimensions are in inches.

MATERIALS: Steel, Corrosion Resisting; QQ-S-765, 60,000 psi minimum tensile strength, 1% minimum elongation in 2 inches.
  Nickel — Copper Alloy (Monel); QQ-N-281, Class A
  Copper; QQ-C-576
  Brass, Half Hard, QQ-B-611, Composition C
  Aluminum Alloy, Half Hard; QQ-A-359

PROTECTIVE COATINGS:
  Corrosion Resisting Steel Washers shall be passivated.
  Aluminum Alloy Washers shall be anodized in accordance with MIL-A-8625 or given a chemical film in accordance with MIL-C-5541.

IDENTIFICATION NUMBER — (MS Number) — (Washer Dash No.)
  Example: MS 15795 — 318 would be the identification number of a ½ nominal size corrosion resisting steel washer with a 1.062 outside diameter.

NOTES: 1. In case of conflict with any referenced document this standard will govern.
    2. Referenced documents shall be the issue in effect on the date at invitation for bids.

APPROVED 2 June 1954 REVISED

CUSTODIANS	OTHER INT.	MILITARY STANDARD	
Ord	A — CESꞮ₆T		
NOrd	N — ASʜYMC	WASHERS, FLAT,	MS15795
USAF	AF —	METAL, ROUND, GENERAL PURPOSE	
PROCUREMENT SPECIFICATION NONE		SUPERSEDES:	SHEET 2 OF 2

HARDWARE

NOMINAL SIZE	INSIDE DIAMETER – A – Max.	INSIDE DIAMETER – A – Min.	WIDTH – W – Min.	THICKNESS $\frac{C=T+t}{2}$ Max.	THICKNESS $\frac{C=T+t}{2}$ Min.	OUTSIDE DIAMETER – B – Max.	STEEL PLAIN (UNCOATED) Dash No.	PLAIN (UNCOATED) FIIN	CADMIUM OR ZINC OPTIONAL Dash No.	CAD/ZINC FIIN	CADMIUM Dash No.	CADMIUM FIIN	PHOSPHATE Dash No.	PHOSPHATE FIIN	CORROSION RESISTING PASSIVATED Dash No.	PASSIVATED FIIN	PHOSPHOR BRONZE CADMIUM Dash No.	CADMIUM FIIN
#2 .086	.097	.088	.030	.021	.015	.165	1	019-2298	20		39		58	019-2303	77	058-2950	96	
#4 .112	.124	.115	.035	.026	.020	.202	2	011-8871	21		40		59	013-1195	78	058-2949	97	
#6 .138	.151	.141	.040	.031	.025	.239	3	011-8872	22		41		60	013-1196	79	043-1754	98	
#8 .164	.178	.168	.047	.037	.031	.280	4	011-8869	23		42		61	013-1197	80	042-9067	99	
#10 .190	.205	.194	.055	.046	.040	.323	5	011-8873	24		43		62	274-8708	81	058-2951	100	
¼	.267	.255	.107	.057	.047	.489	6	011-3114	25		44		63	274-8714	82	043-5862	101	
5⁄16	.333	.319	.117	.066	.056	.575	7	011-2723	26		45		64	013-1201	83		102	
3⁄8	.398	.382	.136	.080	.070	.678	8	011-0730	27		46		65	274-8819	84		103	
7⁄16	.464	.446	.154	.095	.085	.780	9	011-0405	28		47		66	012-1739	85		104	
½	.529	.509	.170	.109	.099	.877	10	010-6500	29		48		67	013-1203	86		105	261-7125
9⁄16	.595	.573	.186	.123	.113	.975	11	011-2724	30		49		68	012-3167	87		106	
5⁄8	.660	.636	.201	.136	.126	1.082	12	010-3334	31		50		69		88		107	189-6811
¾	.791	.763	.233	.163	.153	1.277	13	010-3335	32		51		70	013-1226	89		108	
7⁄8	.922	.890	.264	.199	.179	1.470	14	010-3336	33		52		71	013-1227	90		109	
1	1.053	1.017	.289	.222	.202	1.656	15	011-7661	34		53		72		91		110	
1⅛	1.184	1.144	.314	.244	.224	1.837	16	187-3202	35		54		73	013-1229	92		111	
1¼	1.315	1.271	.336	.264	.244	2.012	17	011-7613	36		55		74		93		112	
1⅜	1.446	1.398	.356	.284	.264	2.183	18	011-8028	37		56		75		94		113	
1½	1.577	1.525	.375	.302	.282	2.352	19	011-8029	38		57		76		95		114	

FED SUP CLASS 5310

PAGE 100

APPENDIX C

Material: Steel, Carbon, FS1060 to FS1080, Rockwell "C" 45-53, Specification QQ-S-633.
Corrosion Resisting Steel, Federal Standard No. 66, Steel Numbers 302 Rockwell "C" 35-43 or 420 Rockwell "C" 43-53.
Phosphor Bronze, Specification QQ-B-746, Composition A, Hard.

Protective Coating:
Cadmium Plate, Specification QQ-P-416, Type II, Class C.
Zinc Plate, Specification QQ-Z-325, Type II, Class 3.
Phosphate, Specification MIL-C-16232, Type II.

Dimensions: All dimensions are in inches unless otherwise specified.
Part Numbers: The MS part number consists of the MS number, plus the dash number. Example: MS35337-1.
Notes:
(1) Referenced documents shall be of the issue in effect on invitations for bid.
(2) This standard takes precedence over documents referenced herein.
(3) This document has been promulgated by the Department of Defense as the Military Standard to limit the selection of the item, product or design covered herein in engineering, design and procurement. This standard shall become effective not later than 90 days after the latest date of approval shown.

APPROVED MAR 4 1954   REVISED 28 APRIL 56

HARDWARE

CUSTODIANS
A – Ord
N – Ships
AF – AF

OTHER INT.
A – CESigT
N – MCOrASY
AF –

MILITARY STANDARD
WASHER, LOCK, SPLIT, HELICAL, LIGHT SERIES

MS35337

PROCUREMENT SPECIFICATION FF-W-84

SUPERSEDES:

SHEET 1 OF 1

SCREW CLEARANCE AND HOLE CHART

SCREW NO.	SCREW BODY DIA.	SINGLE HOLE DIA.	MIN. 82°	CSK. SHT. THICKNESS 100°	CSK. DIA. ±.005
2	.086	.089 (#43)	.063		.151
4	.112	.120 (#31)	.083	.064	.225
6	.138	.144 (#27)	.095	.072	.289
8	.164	.172 (¹¹⁄₆₄)	.109	.081	.337
10	.190	.194 (#11)	.125	.091	.390
¼	.250	.257 (F)	.168	.125	.512
⁵⁄₁₆	.313	.316 (O)	.209	.156	.640
⅜	.375	.386 (W)	.253	.188	.767
⁷⁄₁₆	.438	.453 (²⁹⁄₆₄)	.241	.204	.895
½	.500	.516 (³³⁄₆₄)	.241	.231	1.002
⁹⁄₁₆	.563	.578 (³⁷⁄₆₄)	.271	.251	1.150
⅝	.625	.641 (⁴¹⁄₆₄)	.312	.286	1.277

FOR MULTIPLE HOLE PATTERN USE THE FOLLOWING FORMULAS:*
FOR CLEARANCE HOLE ON TAPPED ONLY

2 HOLE PATTERN $D = d + 2t$
3 HOLE PATTERN $D = d + 4t$
4 HOLE PATTERN $D = d + 2.82t$
6 HOLE PATTERN $D = d + 5.62t$
   OR MORE THAN
   6 HOLES.

WHERE D = CLEARANCE HOLE DIA.
d = SCREW BODY DIA.
t = TOLERANCE (₵ to ₵ HOLE)

FOR CLEARANCE HOLE ON CLEARANCE HOLE
DIVIDE THE LAST PART BY 2

$$\frac{2t}{2} \,,\quad \frac{4t}{2} \quad etc.$$

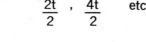

EXAMPLE:

1. What is the hole Dia. for a single #6-32 Binding Head Screw?

   ANSWER: $.144 \,^{+.005}_{-.001}$ DIA.

2. What is the hole Dia. for a single #4-40 82° Flat Head Screw?

   ANSWER: $.120 \,^{+.004}_{-.001}$ DIA. HOLE

   CSINK .225 DIA. x 82°

3. What hole Dia. should be drilled for a #4-40 Binding Head Screw, 4 Places, (Tolerance between holes ₵ to ₵ is ±.005)

   ANSWER; $D = d + 2.82t$
   $D = .112 + (2.82 \times .005)$
   $D = .112 + .0141 = .1261$

   Therefore Hole Dia. should read to the largest Drill Size (See Appendix E)

   $.128 \,^{+.005}_{-.001}$ DIA.

   4 HOLES

STANDARD DRILLED HOLE TOLERANCES	
HOLE DIA.	TOLERANCE
.0135 THRU .125	+.004 −.001
.126 THRU .250	+.005 −.001
.251 THRU .500	+.006 −.001
.501 THRU .750	+.008 −.001
.751 THRU 1.000	+.010 −.001
1.001 THRU 2.000	+.012 −.001

*For exercises see Lesson NO. 1 "MECHANICAL DRAFTING REVIEW."

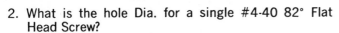

CSK. ANGLE
CSK. DIA.
SHEET THICKNESS
HOLE DIA.

APPENDIX D

SCREW CLEARANCE AND HOLE CHART

FRACTIONS AND DECIMAL EQUIVALENTS

Fraction	Decimal	Fraction	Decimal
1/64	.0156	33/64	.5156
1/32	.0312	17/32	.5312
3/64	.0468	35/64	.5468
1/16	.0625	9/16	.5625
5/64	.0781	37/64	.5781
3/32	.0937	19/32	.5937
7/64	.1093	39/64	.6093
1/8	.125	5/8	.625
9/64	.1406	41/64	.6406
5/32	.1562	21/32	.6562
11/64	.1718	43/64	.6718
3/16	.1875	11/16	.6875
13/64	.2031	45/64	.7031
7/32	.2187	23/32	.7187
15/64	.2343	47/64	.7343
1/4	.25	3/4	.75
17/64	.2656	49/64	.7656
9/32	.2812	25/32	.7812
19/64	.2968	51/64	.7968
5/16	.3125	13/16	.8125
21/64	.3281	53/64	.8281
11/32	.3437	27/32	.8437
23/64	.3593	55/64	.8593
3/8	.375	7/8	.875
25/64	.3906	57/64	.8906
13/32	.4062	29/32	.9062
27/64	.4218	59/64	.9218
7/16	.4375	15/16	.9375
29/64	.4531	61/64	.9531
15/32	.4687	31/32	.9687
31/64	.4843	63/64	.9843
1/2	.5	1	1.

OHM'S LAW

VOLTS(E)	OHMS(R)	AMPERES(I)	WATTS(W)
$E = IR$	$R = E/I$	$I = E/R$	$W = EI$
$E = \sqrt{WR}$	$R = E^2/W$	$I = W/E$	$W = I^2R$
$E = W/I$	$R = W/I^2$	$I = \sqrt{W/R}$	$W = E^2R$

DRILL SIZES — DECIMAL EQUIVALENTS

Drill Size	Decimal Equivalent	Drill Size	Decimal Equivalent	Drill Size	Decimal Equivalent
80	.0135	1/8	.1250	O	.3160
79	.0145	30	.1285	P	.3230
1/64	.0156	29	.1360	21/64	.3281
78	.0160	28	.1405	Q	.3320
77	.0180	9/64	.1406	R	.3390
76	.0200	27	.1440	11/32	.3437
75	.0210	26	.1470	S	.3480
74	.0225	25	.1495	T	.3580
73	.0240	24	.1520	23/64	.3594
72	.0250	23	.1540	U	.3680
71	.0260	5/32	.1562	3/8	.3750
70	.0280	22	.1570	V	.3770
69	.0292	21	.1590	W	.3860
68	.0310	20	.1610	25/64	.3906
1/32	.0313	19	.1660	X	.3970
67	.0320	18	.1695	Y	.4040
66	.0330	11/64	.1719	13/32	.4062
65	.0350	17	.1730	Z	.4130
64	.0360	16	.1770	27/64	.4219
63	.0370	15	.1800	7/16	.4375
62	.0380	14	.1820	29/64	.4531
61	.0390	13	.1850	15/32	.4687
60	.0400	3/16	.1875	31/64	.4843
59	.0410	12	.1890	1/2	.5000
58	.0420	11	.1910	33/64	.5156
57	.0430	10	.1935	17/32	.5312
56	.0465	9	.1960	35/64	.5469
3/64	.0469	8	.1990	9/16	.5625
55	.0520	7	.2010	37/64	.5781
54	.0550	13/64	.2031	19/32	.5937
53	.0595	6	.2040	39/64	.6094
1/16	.0625	5	.2055	5/8	.6250
52	.0635	4	.2090	41/64	.6406
51	.0670	3	.2130	21/32	.6562
50	.0700	7/32	.2187	43/64	.6719
49	.0730	2	.2210	11/16	.6875
48	.0760	1	.2280	45/64	.7031
5/64	.0781	A	.2340	23/32	.7187
47	.0785	15/64	.2344	47/64	.7344
46	.0810	B	.2380	3/4	.7500
45	.0820	C	.2420	49/64	.7656
44	.0860	D	.2460	25/32	.7812
43	.0890	1/4 E	.2500	51/64	.7969
42	.0935	F	.2570	13/16	.8125
3/32	.0937	G	.2610	53/64	.8281
41	.0960	17/64	.2656	27/32	.8437
40	.0980	H	.2660	55/64	.8594
39	.0995	I	.2720	7/8	.8750
38	.1015	J	.2770	57/64	.8906
37	.1040	K	.2811	29/32	.9062
36	.1065	9/32	.2812	59/64	.9219
7/64	.1093	L	.2900	15/16	.9375
35	.1100	M	.2950	61/64	.9531
34	.1110	19/64	.2968	31/32	.9687
33	.1130	N	.3020	63/64	.9844
32	.1160	5/16	.3125	1	1.0000
31	.1200				

COLOR CODE

COLOR	ABBREV.*	NO.	MULTIPLIER	TOL. ±%
BLACK	(BLK) BK	0		
BROWN	(BRN) BR	1	10	
RED	(RED) R	2	100	
ORANGE	(ORN) O	3	1000	
YELLOW	(YEL) Y	4	10^4	CATHODE
GREEN	(GRN) GN	5	10^5	
BLUE	(BLU) BL	6	10^6	ANODE
VIOLET OR PURPLE	(VIO) V (PR) P	7	10^7	
GRAY	(GY) GY	8	10^8	
WHITE	(WHT) W	9	10^9	
GOLD			.1	5%
SILVER			.01	10%
NO COLOR				20%

*ABBREVIATIONS shown in brackets are MILITARY STANDARDS.

*ABBREVIATIONS not shown in brackets were made up specifically for this exercise book.

EXAMPLE: Resistor value by COLOR

- 1st digit — RED — 2
- 2nd digit — YELLOW — 4
- 3rd digit (multiplier) — ORANGE — 3
- 4th digit (tolerance) — GOLD — 5%

Answer; 24000 or 24K, ±5%

(K = 1000, M = 1000000)

LOGIC SYMBOLS

In the following list of logic symbols, explanations accompany the graphic representations.

LOGIC SYMBOL	EXPLANATION

AND

INPUT SIDE   OUTPUT SIDE

A
B ————— F

A
B ————— F
C

The symbol shown represents the AND function.

The AND output is high (H) if and only if all the inputs are high. A **small circle** at the output indicates **an opposite** output from that of the statement and table above. See lower table at right. The small circle shall never be drawn by itself on a diagram.

TRUTH TABLE

H=High		L=Low
Input		Output
A	B	F
L	L	L
L	H	L
H	L	L
H	H	H

AND function with Small Circle

Input			Output
A	B	C	F
L	L	L	H
L	L	H	H
L	H	L	H
L	H	H	H
H	L	L	H
H	L	H	H
H	H	L	H
H	H	H	L

INCLUSIVE OR

INPUT SIDE   OUTPUT SIDE

A
B ————— F

A
B ————— F
C

The symbol shown represents the INCLUSIVE OR function.

The OR output is high (H) if and only if any one or more of the inputs are high.

Inputs can be drawn on any side of the symbol except the output side.

Input		Output
A	B	F
L	L	L
L	H	H
H	L	H
H	H	H

Input			Output
A	B	C	F
L	L	L	L
L	L	H	H
L	H	L	H
L	H	H	H
H	L	L	H
H	L	H	H
H	H	L	H
H	H	H	H

EXCLUSIVE OR

A
B ————— F

The symbol shown represents the EXCLUSIVE OR function.

The EXCLUSIVE OR output is high (H) if and only if any one input is high and all other inputs are low.

Input		Output
A	B	F
L	L	L
L	H	H
H	L	H
H	H	L

FLIP-FLOP

S T C
FF
1   0

Aspect ratio = 1.75 : 1

The flip-flop is a device which stores a single bit of information.

It has three possible inputs, set (S), clear (reset) (C), and toggle (trigger) (T), and two possible outputs, 1 and 0.

When not used, the trigger input may be omitted.

LOGIC SYMBOL	EXPLANATION
BINARY REGISTER 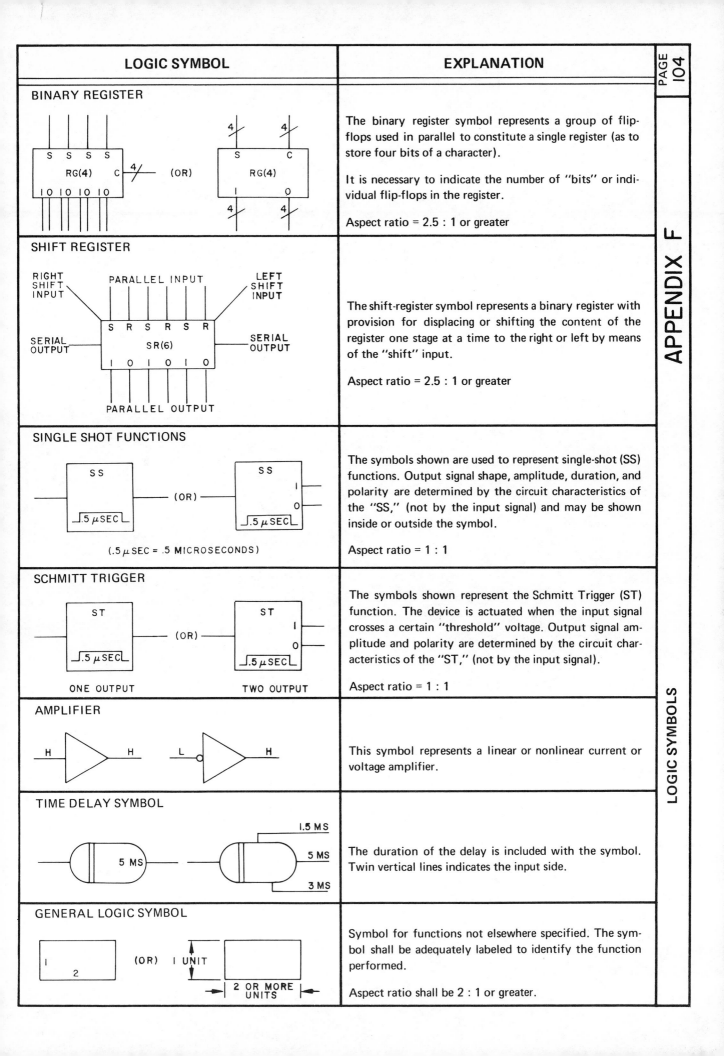	The binary register symbol represents a group of flip-flops used in parallel to constitute a single register (as to store four bits of a character). It is necessary to indicate the number of "bits" or individual flip-flops in the register. Aspect ratio = 2.5 : 1 or greater
SHIFT REGISTER	The shift-register symbol represents a binary register with provision for displacing or shifting the content of the register one stage at a time to the right or left by means of the "shift" input. Aspect ratio = 2.5 : 1 or greater
SINGLE SHOT FUNCTIONS	The symbols shown are used to represent single-shot (SS) functions. Output signal shape, amplitude, duration, and polarity are determined by the circuit characteristics of the "SS," (not by the input signal) and may be shown inside or outside the symbol. Aspect ratio = 1 : 1
SCHMITT TRIGGER	The symbols shown represent the Schmitt Trigger (ST) function. The device is actuated when the input signal crosses a certain "threshold" voltage. Output signal amplitude and polarity are determined by the circuit characteristics of the "ST," (not by the input signal). Aspect ratio = 1 : 1
AMPLIFIER	This symbol represents a linear or nonlinear current or voltage amplifier.
TIME DELAY SYMBOL	The duration of the delay is included with the symbol. Twin vertical lines indicates the input side.
GENERAL LOGIC SYMBOL	Symbol for functions not elsewhere specified. The symbol shall be adequately labeled to identify the function performed. Aspect ratio shall be 2 : 1 or greater.

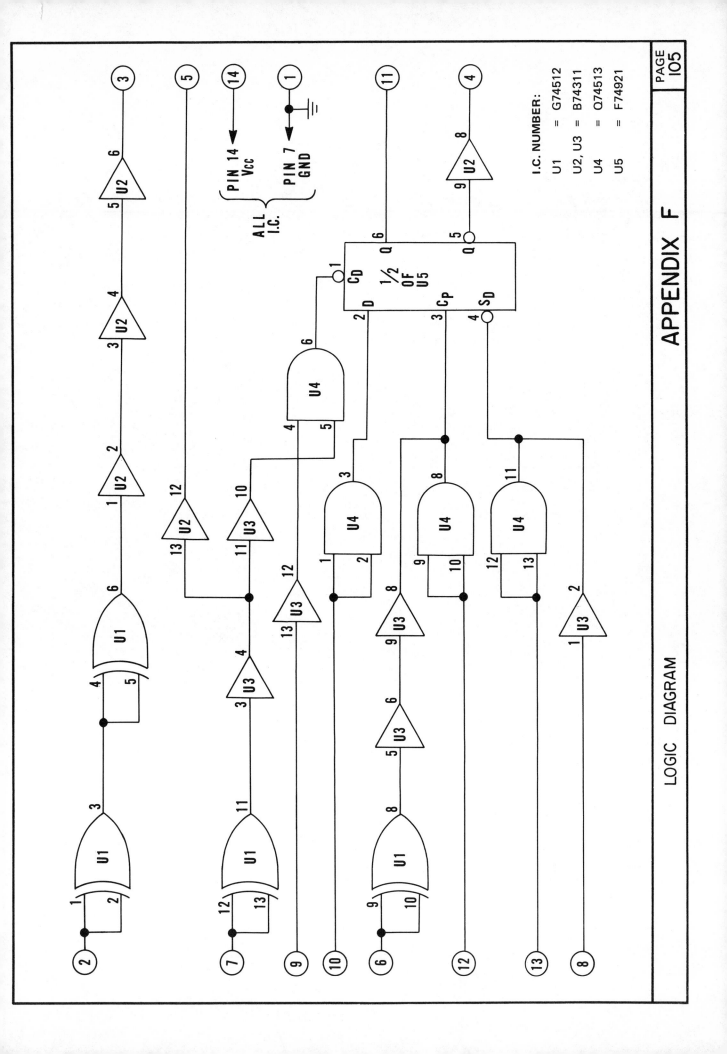

APPENDIX F

LOGIC DIAGRAM

I.C. NUMBER:

U1 = G74512
U2, U3 = B74311
U4 = Q74513
U5 = F74921

FIG. 2 14-PIN DUAL-IN-LINE PACKAGE

0.700

0.250

0.090

0.125

0.020 DIA. LEADS

0.300

0.250

0.100 LEAD SPACING

FIG. 3 I.C. CHIPS

QUAD GATE
Q74513

FLIP-FLOP
F74921

BUFFER
B74311

GATE
G74512

14-PIN DUAL-IN-LINE I.C.

The illustration (Fig. 1) below shows a simple method of mounting a dual-in-line to a printed circuit board.

Pad configurations may vary in shape and O.D. size. The O.D. pad size depends on the pattern of conductor lines crossing on top and bottom of the board.

A common practice is to use the following:

FOR	USE
FULL SCALE layout	.062 O.D. Pad
2 x SIZE layout	.125 O.D. Pad
4 x SIZE layout	.250 O.D. Pad

Due to a wide variation in tolerance of both the diameter and the $\mathcal{C}$ to $\mathcal{C}$ distance of the dual-in-line leads, a .031 dia. hole, as shown, is recommended for ease in mounting the dual-in-line package.

Plated-thru holes are required when the conductor pattern is both on the top and bottom of the P.C.B.

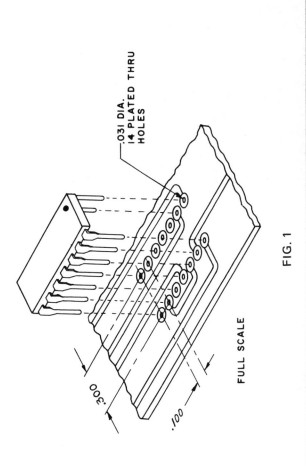

.031 DIA.
14 PLATED THRU HOLES

.300

.100

FULL SCALE

FIG. 1

INTEGRATED CIRCUIT — 14 PINS DUAL–IN–LINE

APPENDIX G

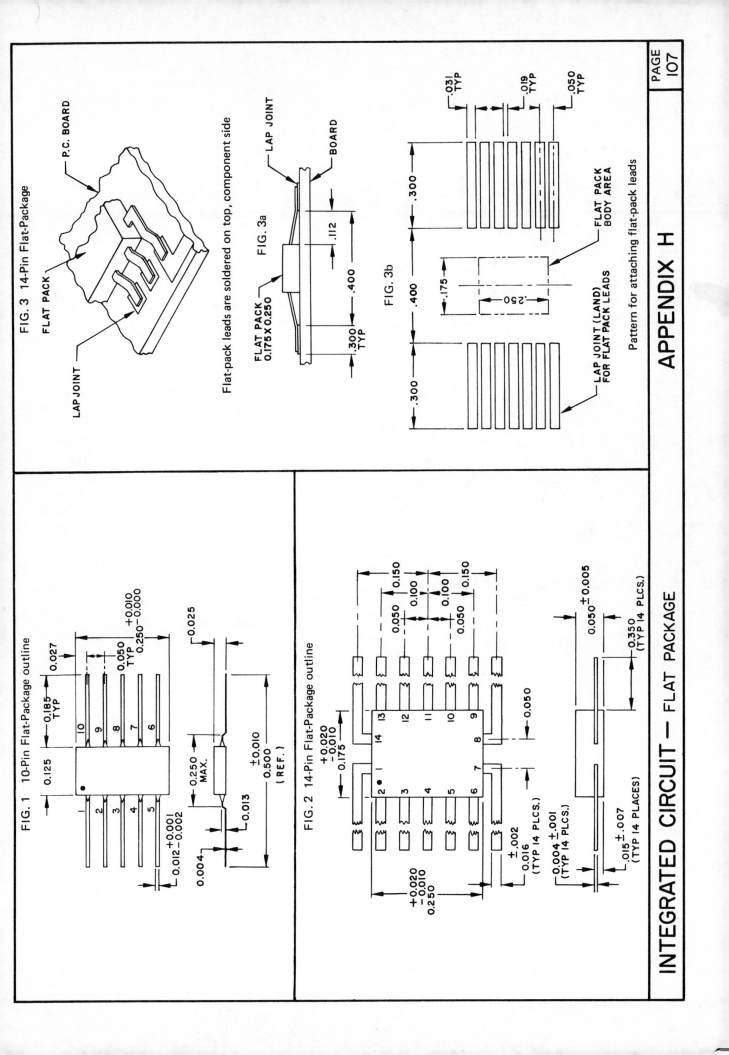

FIG. 3 14-Pin Flat-Package

P.C. BOARD

FLAT PACK

LAP JOINT

Flat-pack leads are soldered on top, component side

FIG. 3a

LAP JOINT

BOARD

FLAT PACK
0.175 x 0.250

.112

.400

.300 TYP

FIG. 3b

.300

.400

.175

.250

.300

.031 TYP

.019 TYP

.050 TYP

FLAT PACK BODY AREA

LAP JOINT (LAND) FOR FLAT PACK LEADS

Pattern for attaching flat-pack leads

FIG. 1 10-Pin Flat-Package outline

0.027

0.185 TYP

0.125

0.050 TYP

+0.010
0.250 −0.000

0.025

0.250 MAX.

±0.010
0.500
(REF.)

0.013

0.012 +0.001 −0.002

0.004

FIG. 2 14-Pin Flat-Package outline

+0.020
−0.010
0.175

+0.020
−0.010
0.250

0.150

0.100

0.050

0.100

0.050

0.150

0.050

0.050 ±0.005

0.350
(TYP 14 PLCS.)

± .002
0.016
(TYP 14 PLCS.)

0.004 ±.001
(TYP 14 PLCS.)

.015 ±.007
(TYP 14 PLACES)

APPENDIX H

PAGE
107

INTEGRATED CIRCUIT — FLAT PACKAGE